Solving Problems in Point Geometry
Insights and Strategies for Mathematical Olympiad and Competitions

Mathematical Olympiad Series

ISSN: 1793-8570

Series Editors: Lee Peng Yee *(Nanyang Technological University, Singapore)*
Xiong Bin *(East China Normal University, China)*

Published

Vol. 23 *Solving Problems in Point Geometry:*
Insights and Strategies for Mathematical Olympiad and Competitions
by Jingzhong Zhang (Guangzhou University, China &
Chinese Academy of Sciences, China),
Xicheng Peng (Central China Normal University, China)

Vol. 22 *Mathematical Olympiad in China (2021–2022):*
Problems and Solutions
editor-in-chief Bin Xiong (East China Normal University, China)

Vol. 21 *Problem Solving Methods and Strategies in High School*
Mathematical Competitions
edited by Bin Xiong (East China Normal University, China) &
Yijie He (East China Normal University, China)

Vol. 20 *Hungarian Mathematical Olympiad (1964–1997):*
Problems and Solutions
by Fusheng Leng (Academia Sinica, China),
Xin Li (Babeltime Inc., USA) &
Huawei Zhu (Shenzhen Middle School, China)

Vol. 19 *Mathematical Olympiad in China (2019–2020):*
Problems and Solutions
edited by Bin Xiong (East China Normal University, China)

Vol. 18 *Mathematical Olympiad in China (2017–2018):*
Problems and Solutions
edited by Bin Xiong (East China Normal University, China)

Vol. 17 *Mathematical Olympiad in China (2015–2016):*
Problems and Solutions
edited by Bin Xiong (East China Normal University, China)

The complete list of the published volumes in the series can be found at
http://www.worldscientific.com/series/mos

Vol. 23 | Mathematical Olympiad Series

Solving Problems in Point Geometry

Insights and Strategies for Mathematical Olympiad and Competitions

Authors

Jingzhong Zhang
Guangzhou University, China & Chinese Academy of Sciences, China

Xicheng Peng
Central China Normal University, China

Translators

Yongsheng Rao, Siran Lei, Ying Wang
Guangzhou University, China

Proofreader

Yongming Liu
East China Normal University, China

Copy Editors

Lingzhi Kong, Liyu Zhang, Ming Ni
East China Normal University Press, China

Published by

East China Normal University Press
3663 North Zhongshan Road
Shanghai 200062
China

and

World Scientific Publishing Co. Pte. Ltd.
5 Toh Tuck Link, Singapore 596224
USA office: 27 Warren Street, Suite 401-402, Hackensack, NJ 07601
UK office: 57 Shelton Street, Covent Garden, London WC2H 9HE

Library of Congress Control Number: 2024033296

British Library Cataloguing-in-Publication Data
A catalogue record for this book is available from the British Library.

Mathematical Olympiad Series — Vol. 23
SOLVING PROBLEMS IN POINT GEOMETRY
Insights and Strategies for Mathematical Olympiad and Competitions

Copyright © 2025 by East China Normal University Press and
World Scientific Publishing Co. Pte. Ltd.

All rights reserved. This book, or parts thereof, may not be reproduced in any form or by any means, electronic or mechanical, including photocopying, recording or any information storage and retrieval system now known or to be invented, without written permission from the Publisher.

For photocopying of material in this volume, please pay a copying fee through the Copyright Clearance Center, Inc., 222 Rosewood Drive, Danvers, MA 01923, USA. In this case permission to photocopy is not required from the publisher.

ISBN 978-981-12-9410-5 (hardcover)
ISBN 978-981-12-9475-4 (paperback)
ISBN 978-981-12-9411-2 (ebook for institutions)
ISBN 978-981-12-9412-9 (ebook for individuals)

For any available supplementary material, please visit
https://www.worldscientific.com/worldscibooks/10.1142/13865#t=suppl

Typeset by Stallion Press
Email: enquiries@stallionpress.com

Preface

Elementary geometry is more than two thousand years old, dating back to the time of Euclid. The difficulty of learning it lies in the construction of ever-changing auxiliary graphics. Mastering the creation of auxiliary graphics takes a lot of time. When you come across a new problem, you often feel stuck. If the coordinate system is established and solved by analytic geometry methods, more calculations are needed. So, Leibniz asked, can we calculate the points directly and find new ways to solve geometric problems more efficiently?

We try to answer Leibniz's question in a new way: by building a geometric system that operates directly on points, so we call it "point geometry". This method combines the advantages of the coordinate method, vector method, and mass point geometry method, and the operations are simpler. This book begins with the basic concepts and algorithms of point geometry, from easy to difficult, and discusses in detail the methods of solving elementary geometry problems with point geometry. The book not only lists the key points of point geometry but also discusses the relationships among the point geometry method, vector method, analytic method, mass point method.

It is worth paying special attention to our proposed identity method. Using this method, we can often prove difficult competition problems in equational geometry using a single line of equations, easily find any superfluous conditions in the problems, and obtain multiple propositions from this one line of equations. I believe readers can revel in the joy of mathematical thinking from the solutions to several examples in the book.

In connection with the development of artificial intelligence, the point geometry method may have more important significance. Here, we found a relatively simple and geometric meaning of rich knowledge representation, as well as a concise and bright way of reasoning, through which the proofs of many geometric propositions become simple! This makes it easy to achieve mechanization. In fact, most of the answers to the questions in this book are mechanically generated by computer programs, and the author has only provided the necessary tests and explanations. Can we deepen and generalize this line of thinking so that more knowledge representations are made simple and meaningful and that relevant reasoning is not only simple and easy to implement but also efficient and easily understandable? This is a very significant and arduous task. If it can be successful, even partially, in a certain field, it can greatly simplify the relevant reasoning methods, make people see the relationships between things more clearly, greatly reduce the burden of teaching and learning, and contribute to the discovery of new knowledge. The automatic reasoning field of artificial intelligence will be more colorful and move forward faster.

About the Authors

Jingzhong Zhang, the academician of the Chinese Academy of Sciences, is currently a professor in Guangzhou University. He is a famous Chinese mathematician and computer expert, serves as the chairman of the International Association for Educational Mathematics, and was the chairman of the China Science Writers Association. He is mainly engaged in the research of mechanical theorem proving in geometry, educational information technology, educational mathematics, and is devoted to the creation of popular science in mathematics. He has proposed a new Geometric Algebra, Point Geometry. He has received seven important national awards, including the State Natural Science Award of China, the State Science and Technology Advancement Award of China, and the National Teaching Achievement Award of China.

Xicheng Peng, who works at Central China Normal University, is passionate about popularizing mathematics through written discourse. He has authored over ten works in popular science in mathematics. In recent years, he has been engaged in the research of mechanical theorem proving in Point Geometry and educational mathematics, proposed algorithms for point geometry identity equations and complex geometry identity equations.

Contents

Preface v

About the Authors vii

1. **Overview of Point Geometry** 1

    1.1 Introduction 1
    1.2 Addition and Scalar Multiplication of Points 2
    1.3 Inner Product of Points 5
    1.4 Exterior Product of Points 8
        1.4.1 Exterior product of two points 8
        1.4.2 Exterior product of three points 10
    1.5 Points Multiplication by Complex Number 15
    1.6 Conclusion 19

2. **Computational Methods** 21

    2.1 Intersections Computing and Geometric
        Constructions 21
    2.2 Point-Line Positions 27
    2.3 Perpendicular and Equal Segments 37
    Reference 59

3. **Identity-Based Method 1: Analytical Approaches** 61

    3.1 Classic Vector Identities 61
    3.2 Examples: Identities with Two Terms 66
    3.3 Examples: Identities with Multiple Terms 73

4. Identity-Based Method 2: Undetermined Coefficients   95

    4.1   Method of Undetermined Coefficients 95
    4.2   Applications 112

5. Assorted Problems   145

    5.1   Introducing Parameters 145
    5.2   Introducing Complex Numbers 151
    5.3   Combined with Other Methods 161
    5.4   Trajectory Problems 167

Appendix   Exercise Answers   171

    A.1   Exercise 2.2 171
    A.2   Exercise 2.3 173
    A.3   Exercise 3.2 195
    A.4   Exercise 3.3 195
    A.5   Exercise 4.1 200
    A.6   Exercise 4.2 207
    A.7   Exercise 5.1 220
    A.8   Exercise 5.2 220
    A.9   Exercise 5.3 223
    A.10   Exercise 5.4 223

Chapter 1
Overview of Point Geometry

1.1 Introduction

Using the coordinate method to solve geometric problems offers the advantage of employing systematic algebraic techniques. However, representing geometric points using coordinates can make both the notation and comprehension more intricate, leading to a decrease in intuitive understanding. Additionally, the geometric significance often remains obscured during algebraic calculations. In light of this, Leibniz posed a question: Can we perform calculations directly on geometric objects?

The advent of vector geometry can be seen as an initial response to Leibniz's query. In this direction, mathematicians forged the realm of "geometric algebra" and undertook in-depth investigations. Recent noteworthy work in this field includes the remarkable accomplishments of Li Hongbo regarding conformal geometric algebra.

To overcome certain shortcomings of vector geometry while preserving its advantages, the eminent mathematical logician Mo Shaokui proposed a theory and methodology for mass point geometry that holds greater physical significance. Similarly, Yang Xuezhi introduced point magnitudes, further enriching the theory and methods of geometric algebra. These contributions offer more intuitive, simpler, and easier-to-learn concepts, notations, and forms of operations.

"Point geometry", introduced in this book, strives to describe relationships between geometric objects in a more concise and everyday manner, employing algebraic operations. Point geometry retains the simplicity and intuitiveness of mass point geometry, requiring fewer prerequisites and restrictions on operational conditions while having a wider scope of applicability.

1.2 Addition and Scalar Multiplication of Points

Let capital Latin letters represent points, and lowercase Greek or Latin letters represent real numbers.

In secondary school, we learned about the number line and further discussed the Cartesian coordinate system, which associates points with numbers or ordered pairs. This naturally leads to a question: Can points be added like numbers?

On the number line, if $A = 2$, $B = 5$, and $C = 7$, can we say $A + B = C$? At first glance, it seems correct, but upon closer examination, there's an issue: if we move the origin one unit to the right, then $A = 1$, $B = 4$, and $C = 6$, and the equation $A + B = C$ no longer holds.

However, if $A = 2$, $B = 4$, and $C = 6$, then we have the equation $A - B = B - C$, which is equivalent to $A + C = 2B$. In this case, no matter how we move the origin, we always have $A + C = 2B$. This equation describes the geometric fact that B is the midpoint of segment AC', and it is independent of coordinates.

In general, when representing linear equations between point coordinates as the same relationships between the points themselves, the resulting equations fall into two categories: those that remain unchanged under coordinate transformations and those that change. Clearly, equations that remain unchanged under coordinate transformations have the property that the sum of the coefficients on both sides is equal. In the study of mass point geometry or point magnitudes, only these types of equation relationships are discussed.

However, when $A = 2$, $B = 5$, and $C = 7$, the equation $A + B = C$ does indeed describe a mathematical fact. As long as we do not change the origin in the process of discussing the problem, it is always valid.

Upon further reflection, the equation $A + B = C$ actually describes a geometric relationship between four points, including the origin O, i.e., $A + B = C + O$. Its geometric meaning is "the midpoints of line segments AB and OC coincide". In mass point geometry, it is considered a drawback that it relies on the origin and is not an invariant under coordinate transformations; therefore, it is neither acknowledged nor discussed.

From another perspective, if the equation $A + B = C$ describes the relationship between four points using only three letters, can we utilize its simplicity and elegance? Partially based on the above considerations, the following basic operations are introduced in plane point geometry:

Definition 1. Point addition. If $A = (x_A, y_A)$, $B = (x_B, y_B)$, and $C = (x_A + x_B, y_A + y_B)$, then we denote $A + B = C$.

Definition 2. Scalar multiplication. If $A = (x, y)$ and $B = (\lambda x, \lambda y)$, then we denote $B = \lambda A$.

Clearly, the above operations depend on the choice of the coordinate origin O. Essentially, there are no new concepts here; it's just a simple notation for representing calculations between point coordinates. Using A to represent (x_A, y_A) has its advantages: not only does it reduce the writing effort for points to one-seventh of the coordinate writing effort, but it also simplifies the visualization, which is conducive to geometric intuition and logical thinking.

Obviously, the various laws satisfied by these two operations can inherit relevant parts of the laws of real number operations. The generalization to three- and higher-dimensional spaces is trivial.

Using the notation established by the above conventions, we can immediately deduce the geometric meanings of the following basic operational formulas:

Property 1. If $B = \lambda A$, then O, A, and B are collinear, and $\overrightarrow{OB} = \lambda \overrightarrow{OA}$.

Property 2. If $A + B = \lambda P$, then for different values of λ, point P occupies different positions, as shown in Figure 1.1:

1. If $A + B = C$, then $AOBC$ is a parallelogram.
2. If $A + B = 2M$, then M is the midpoint of AB.
3. If $A + B = 3P$, then P is the centroid of $\triangle OAB$.

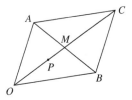

Figure 1.1

Note. In mass point geometry, only $A + B = 2M$ is possible.

Property 3. The difference between two points is a vector, i.e., $B - A = tP = \overrightarrow{AB}$. When $t = 1$, $OABP$ is a parallelogram.

Property 4. The significance of the linear combination of two points, $uA + vB = tP$, is as follows: (1) When $t = u + v$, let $uA + vB = (u+v)F$, which can be rewritten as $u(A - F) = v(F - B)$. It is evident that in this case, point F lies on the line AB, and $\frac{\overrightarrow{AF}}{\overrightarrow{FB}} = \frac{v}{u}$. Its geometric meaning is the same as in particle geometry. (2) Generally, point P lies on the line OF.

Note. Although the geometric facts described by the equation $uA + vB = (u+v)F$ are the same as in mass point geometry, the meanings of terms like uA, vB, and $(u+v)F$ are different here compared to mass point geometry. In mass point geometry, when $u \neq 1$, uA and A represent points with the same position but different masses. However, in point geometry, if $u \neq 1$ and A is not the origin, they are two distinct points with different positions.

For instance, the geometric significance of the equation $2A + 3B = 5F$ is illustrated in Figure 1.2, where $C = 2A$, $D = 3B$, and $E = 5F$, and quadrilateral $OCED$ is a parallelogram. Alternatively, it can be stated that in parallelogram $CODE$, if we draw the line segment AB connecting the midpoint of CO, point A, and the trisection point of OD, point B, and it intersects the diagonal OE at F, then $OE = 5OF$ and $2AF = 3FB$. It is evident that the geometric interpretation of the same equation is more profound in point geometry.

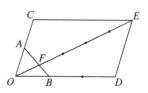

Figure 1.2

Property 5. The representation of the intersection of two lines as $uA + vB = rC + sD$: (1) When $u + v = r + s \neq 0$, let $(u+v)F = uA + vB = rC + sD = (r+s)G$. Then, the equation represents that $F = G$ is the intersection point of lines AB and CD (akin to mass point geometry). (2) In general, the situation represents that O, G, and F are collinear. (3) In particular, if D is the origin, the intersection of AB and CD at F can be represented as $(u+v)F = uA + vB = rC$, where r can be any nonzero real number.

The advantage of point geometry lies in the use of a concise set of symbols with clear meanings to faithfully depict geometric facts, thereby reducing mental effort. This can be exemplified by the following examples.

Example 1. Prove that the medians AM, BN, and CP of $\triangle ABC$ intersect at a point (Figure 1.3).

Proof. Taking A as the origin, let G be the intersection point of BN and CP. Given the conditions, we have $B = 2P$ and $C = 2N$, which implies

$$2M = B + C = 2P + C = 2N + B = 3G.$$

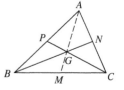

Figure 1.3

This equation indicates that G lies on AM, and $3\overrightarrow{AG} = 2\overrightarrow{AM}$.

Example 2. Let M be the midpoint of side AB of parallelogram $ABCD$, and connect DM intersecting the diagonal AC at P. Prove that $AC = 3AP$ (Figure 1.4).

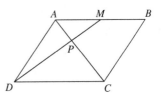

Figure 1.4

Proof. Taking A as the origin, from the given conditions, we have $2M = B = C - D$, which implies $C = 2M + D = 3P$, and hence $\overrightarrow{AC} = 3\overrightarrow{AP}$.

The above solution can be easily transformed into a vector method: from $2M = B = C - D$, we know that $2\overrightarrow{AM} = \overrightarrow{AB} = \overrightarrow{DC}$, which means $2(\overrightarrow{AP} + \overrightarrow{PM}) = \overrightarrow{DC} = \overrightarrow{DP} + \overrightarrow{PC}$. Applying the fundamental theorem of plane vectors, we get $2\overrightarrow{AP} = \overrightarrow{PC}$, and thus $\overrightarrow{AC} = 3\overrightarrow{AP}$.

Example 3. As shown in Figure 1.5, it is known that $AC = 3AP$ and $3AB = 5AQ$. Determine the values of $\frac{PD}{BD}$, $\frac{QD}{CD}$, $\frac{PQ}{RQ}$, and $\frac{BR}{CR}$.

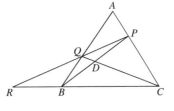

Figure 1.5

Solution: Taking A as the origin, from the given conditions, we have $C = 3P$ and $5Q = 3B$.

Adding the two equations yields $C + 5Q = 3P + 3B = 6D$, which gives $\frac{\overrightarrow{PD}}{\overrightarrow{DB}} = 1$ and $\frac{\overrightarrow{QD}}{\overrightarrow{DC}} = \frac{1}{5}$.

Subtracting the two equations gives $3B - C = 5Q - 3P = 2R$, which results in $\frac{\overrightarrow{BR}}{\overrightarrow{CR}} = \frac{1}{3}$ and $\frac{\overrightarrow{PR}}{\overrightarrow{QR}} = \frac{5}{3}$; hence, $\frac{\overrightarrow{PQ}}{\overrightarrow{QR}} = \frac{2}{3}$.

1.3 Inner Product of Points

Usually, a product is the result of multiplication, and multiplication refers to an operation that distributes over addition. For addition in point geometry, there is more than one operation that distributes, and the most fundamental one is the inner product.

Definition 3. Inner Product. If in a Cartesian coordinate system, $A = (x, y)$ and $B = (u, v)$, then we define

$$A \cdot B = ux + vy.$$

It is evident that the inner product of two points is a real number that depends on the coordinate origin.

According to the definition of inner product, we clearly have the following properties.

Property 6. $A \cdot B = B \cdot A$, and the inner product distributes over addition.

It is easy to verify the following property.

Property 7. $(A - B) \cdot (C - D) = \overrightarrow{AB} \cdot \overrightarrow{CD}$.

Let $A^2 = A \cdot A$, then $(A - B)^2 = |\overrightarrow{AB}|^2$, and A^2 is the square of the distance from point A to the origin.

When both A and B are not the origin, $A \cdot B = 0$ implies that $\angle AOB$ is a right angle; $(A - B) \cdot (C - D) = 0$ implies that $\overrightarrow{AB} \perp \overrightarrow{CD}$.

Example 1. Prove that the altitudes AD, BE, and CF of $\triangle ABC$ are concurrent.

Proof. As shown in Figure 1.6, take the intersection point H of AD and BE as the origin. The given conditions imply $(B - C) \cdot A = 0$ and $(A - C) \cdot B = 0$. Subtracting the two equations yields $(B - A) \cdot C = 0$, which proves that $\overrightarrow{CH} \perp \overrightarrow{AB}$.

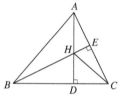

Figure 1.6

Example 2. Given $\triangle ABC$ with sides a, b, and c, find the length of median AD and the angle of AE.

Solution. (1) For median AD: take A as the origin, then the square of median AD is D^2. Using the equation $B + C = 2D$, we get $B^2 + C^2 + 2B \cdot C = 4D^2$. To eliminate the $B \cdot C$ term, use the equation $(B + C)^2 - (B - C)^2 = 4B \cdot C$, which gives

$$2B \cdot C = \frac{(B+C)^2}{2} - \frac{(B-C)^2}{2} = 2D^2 - \frac{a^2}{2}.$$

Substitute this into the previous equation, we have $c^2 + b^2 + 2D^2 - \frac{a^2}{2} = 4D^2$, and solving for $|AD|$ yields

$$|AD| = \frac{\sqrt{2(b^2 + c^2) - a^2}}{2}.$$

(2) For angle bisector AE: take A as the origin. The square of the angle bisector AE is E^2. Using $\frac{|BE|}{|CE|} = \frac{c}{b}$, we have $cC + bB = (b+c)E$. Squaring

this equation gives
$$c^2C^2 + b^2B^2 + 2bcB \cdot C = (b+c)^2 E^2.$$

To eliminate the $B \cdot C$ term, we use the equation previously derived for the median:

$$2B \cdot C = \frac{(B+C)^2}{2} - \frac{(B-C)^2}{2} = 2D^2 - \frac{a^2}{2} = b^2 + c^2 - a^2.$$

Substituting this into the equation above, we get $2b^2c^2 + bc(b^2 + c^2 - a^2) = (b+c)^2 E^2$, and solving for $|AE|$ yields

$$|AE| = \sqrt{bc\left(1 - \frac{a^2}{(b+c)^2}\right)}.$$

Example 3. In Figure 1.7, let H be the orthocentre of $\triangle ABC$. Points D, E, and F are midpoints of sides BC, CA, and AB, respectively. Points L, M, and N are on lines BC, CA, and AB, respectively, such that AL, BM, and CN are perpendicular to DH, EH, and FH, respectively. Prove that the points L, M, and N are collinear, and the line is perpendicular to the Euler line.

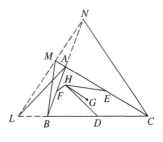

Figure 1.7

Solution. Take the orthocentre H of $\triangle ABC$ as the origin. Then, we have

$$A \cdot (B - C) = B \cdot (A - C) = C \cdot (A - B) = 0.$$

Let G be the centroid of $\triangle ABC$, and let D, E, and F be the midpoints of sides BC, CA, and AB, respectively. We have

$$A + B + C = 3G, \quad B + C = 2D, \quad C + A = 2E, \quad A + B = 2F.$$

Let line L intersect line BC such that $AL \perp DH$, and let point M be intersected on line AC such that $BM \perp EH$. Then,

$$L = uB + (1-u)C, \quad M = vA + (1-v)C,$$
$$(A - L) \cdot D = 0, \quad (B - M) \cdot E = 0.$$

We need to prove that $ML \perp GH$, which is equivalent to proving $(M-L) \cdot G = 0$. From $(A-L) \cdot D = 0$, we have

$$\begin{aligned}
0 &= (A - uB - (1-u)C) \cdot (B+C) \\
&= (A - C - u(B-C)) \cdot (B+C) \\
&= (A-C) \cdot C - u(B-C) \cdot (3G - A) \\
&= (A-C) \cdot C - 3u(B-C) \cdot G \text{ (eliminating } D\text{).}
\end{aligned}$$

Similarly, from $(B-M) \cdot E = 0$, we have

$$\begin{aligned}
0 &= (B - vA - (1-v)C) \cdot (A+C) \\
&= (B-C) \cdot C - 3v(A-C) \cdot G \text{ (eliminating } E\text{).}
\end{aligned}$$

Hence, we get

$$\begin{aligned}
3(M-L) \cdot G &= 3(v(A-C) - u(B-C)) \cdot G \\
&= (B-C) \cdot C - (A-C) \cdot C \\
&= (B-A) \cdot C = 0.
\end{aligned}$$

Thus, the collinearity is proven.

Note. The process in this proof is rather lengthy using the computational method. For a shorter proof, refer to Example 28 in Section 4.2.

1.4 Exterior Product of Points

1.4.1 *Exterior product of two points*

Definition 4. Exterior Product of Two Points. It is conventionally defined that the exterior product of two points, A and B, is $AB = B - A$.

This indicates that the exterior product of two points is the difference between the two points, which is a vector.

This might seem a bit odd. Why call the difference a product? Is it merely to save a minus sign? Further exploration of the properties of the exterior product of two points will reveal the advantages of this definition.

According to the definition, the following properties are evident.

Property 8. $AB = -BA$.

Property 9. If $uA + vB = (u+v)C$, then we have

$$uAP + vBP = (u+v)CP,$$
$$uPA + vPB = (u+v)PC.$$

The correctness of these two equations can be verified by expanding based on the definition. This can be seen as the distributive property of the exterior product over addition. Following this rule, if we apply the exterior product to both sides of the equation $uA + vB = (u+v)C$ using B, we get $uAB = (u+v)CB$; applying it using C, we get $uAC + vBC = 0$, that is, $uAC = vCB$.

However, in general, when the sum of coefficients on both sides of the equation is not equal, the distributive property might not hold, and a correction term needs to be added.

Property 10. If $uA + vB = rC$, considering that the origin O has zero coordinates, we have $uA + vB = rC + (u + v - r)O$. In this case, the distributive property can be applied. Notably, if we observe that $P = OP = -PO$, we get

$$uAP + vBP = rCP + (u+v-r)P,$$
$$uPA + vPB = rPC - (u+v-r)P.$$

The correctness of these two equations can also be verified by expanding based on the definition.

This can be viewed as a generalization of the distributive property of the exterior product over addition. When $u + v = r$, we obtain Property 9. As a special case, we can also deduce the following: (1) If $A = rC$, then $AP = rCP + (1-r)P$ and $PA = rPC - (1-r)P$. (2) If $uA = rC$, then $uAP = rCP + (u-r)P$ and $uPA = rPC - (u-r)P$.

Note. The above laws can be extended to the case of sums of multiple terms. The approach remains the same: directly apply the distributive property when the sum of coefficients on both sides is equal; otherwise, add an origin term to balance the sum of coefficients and then apply the distributive property.

Example 1. Let $ABCD$ be a parallelogram, E be the intersection point of its diagonals, and P be any point. Given $\overrightarrow{PA}$, $\overrightarrow{PB}$, and $\overrightarrow{PC}$, find $\overrightarrow{PD}$ and $\overrightarrow{PE}$.

Solution. Take B as the origin. Then, from the given conditions, we have $A + C = 2E = D$. Therefore,
$$PA + PC = 2PE = PD - P,$$
that is, $\overrightarrow{PA} + \overrightarrow{PC} = 2\overrightarrow{PE} = \overrightarrow{PD} - \overrightarrow{BP}$. The rest of the solution is omitted.

1.4.2 Exterior product of three points

Definition 5. Exterior Product of Three Points. The exterior product of three points A, B, and C, denoted as ABC, represents the signed area of $\triangle ABC$.

Specifically, if $A = (x_A, y_A)$, $B = (x_B, y_B)$, and $C = (x_C, y_C)$, then the exterior product ABC is given by
$$ABC = \frac{1}{2}(x_A y_B + x_B y_C + x_C y_A - x_A y_C - x_B y_A - x_C y_B).$$

From this definition, we have the following properties.

Property 11. $ABC = BCA = CAB = -ACB = -BAC = -CBA$.

The geometric interpretation is intuitive: in a right-handed coordinate system, if the vertices A, B, and C of $\triangle ABC$ are arranged anticlockwise, then $ABC > 0$; otherwise, $ABC < 0$. If A, B, and C are collinear, then $ABC = 0$.

Property 12. If $uA + vB = (u + v)C$, then
$$uAPQ + vBPQ = (u + v)CPQ,$$
and as a result,
$$uPAQ + vPBQ = (u + v)PCQ,$$
$$uPQA + vPQB = (u + v)PQC.$$

The correctness of these equalities can be verified directly. This demonstrates that when the coefficients on both sides of an equality sum up to the same value, the exterior product of three points adheres to the distributive law.

When dealing with $u + v = r + s \neq 0$, the equality $(u + v)F = uA + vB = rC + sD$ represents F as the intersection of lines AB and CD. Applying the exterior product to CD gives $uACD + vBCD = 0$, implying $\frac{ACD}{BCD} = -\frac{v}{u} = \frac{\overrightarrow{AF}}{\overrightarrow{BF}}$, which is known as the "co-side theorem".

Note. The co-side theorem proves valuable when handling problems with multiple variable parameters, aiding in parameter reduction and simplifying calculations, as exemplified in the third proof of Example 3 later.

Property 13. If $uA + vB = rC$, with the origin at O, then

$$uAPQ + vBPQ = rCPQ + (u + v - r)OPQ.$$

This signifies that when the coefficients on both sides of an equality are not equal, an additional balancing term can be introduced to equate the coefficients, followed by the application of the distributive law. An instructive special case emerges when the origin lies on line PQ; in this instance, the additional balancing term becomes zero, effectively ensuring the validity of the distributive law.

Example 2. Given $\triangle ABC$ with points P, Q, and R chosen on its sides such that $BP = PC$, $CQ = 2QA$, and $AR = 3RB$. The lines AP, BQ, and CR form $\triangle LMN$, as shown in Figure 1.8. Determine the value of $\frac{S_{\triangle LMN}}{S_{\triangle ABC}}$.

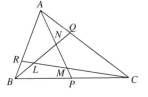

Figure 1.8

Solution. Let A be the origin. Then, from the given conditions, we have

$$B + C = 2P, \tag{1}$$
$$C = 3Q, \tag{2}$$
$$3B = 4R. \tag{3}$$

Substituting (2) into (1), we get $2P = B + 3Q = 4N$. Substituting (3) into (1), we get $6P = 4R + 3C = 7M$. Multiplying (3) by 2 and adding it to (2), we get

$$C + 8R = 3Q + 6B = 9L.$$

Expressing L, M, and N in terms of A, B, and C and balancing coefficients, we have

$$9L = 3Q + 6B = C + 6B = 2A + 6B + C,$$
$$7M = 4R + 3C = 3B + 3C = A + 3B + 3C,$$
$$4N = B + 3Q = B + C = 2A + B + C.$$

Taking the exterior products of these three equalities and omitting terms with zero values, we obtain

$$252LMN = 6ABC + 6ACB + 6BAC + 36BCA$$
$$+ CAB + 6CBA = 25ABC.$$

Therefore, $\frac{S_{\triangle LMN}}{S_{\triangle ABC}} = \frac{25}{252}$.

Example 3 (Pappus's theorem). Let points A, B, and C be collinear and points D, E, and F be collinear. Lines AE and BD intersect at P, lines AF and CD intersect at Q, and lines CE and BF intersect at R, as shown in Figure 1.9.

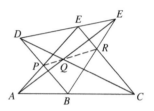

Figure 1.9

Proof 1. Taking P as the origin, assume

$$E = uA, \quad B = vD, \tag{1}$$

$$rA + (1-r)B = C, \tag{2}$$

$$sD + (1-s)E = F. \tag{3}$$

To obtain the expressions for points Q and R, let t and x be undetermined parameters. Multiply (2) by t, and add to (3):

$$trA + t(1-r)B + sD + (1-s)E = tC + F. \tag{4}$$

Multiply (2) by x, and add to (3):

$$xrA + x(1-r)B + sD + (1-s)E = xC + F. \tag{5}$$

Using (1), eliminate B and E in (4), and eliminate A and D in (5), yielding

$$trA + tv(1-r)D + sD + u(1-s)A = tC + F, \tag{6}$$

$$\frac{xr}{u}E + x(1-r)B + \frac{s}{v}B + (1-s)E = xC + F. \tag{7}$$

Rewrite (6) and (7) to obtain the expressions for Q and R:

$$(tr + u(1-s))A - F = tC - (tv(1-r) + s)D = (tr + u(1-s) - 1)Q, \tag{8}$$

$$v(xr + u - us)E - xuvC = uvF - u(xv(1-r) + s)B$$
$$= v(xr + u - us - xu)R. \tag{9}$$

It should hold that $tr + u(1-s) - 1 = t - (tv(1-r) + s)$, solving for t:
$$t = \frac{(1-s)(1-u)}{(v-1)(1-r)}.$$
Similarly, it should hold that $v(xr+u-us)-xuv = uv-u(xv(1-r)+s)$, solving for x:
$$x = \frac{us(v-1)}{vr(1-u)}.$$
Using (1), (2), and (3) to express the left-hand sides of (8) and (9) as combinations of A and D, respectively,
$$u^2(1-s)A - xuv(1-r)D = (xr + u - us - xu)R, \tag{10}$$
$$trA - sD = (tr + u(1-s) - 1)Q. \tag{11}$$
It can be easily verified that $\frac{u^2(1-s)}{tr} = \frac{u^2(v-1)(1-r)}{r(1-u)} = \frac{xuv(1-r)}{s}$, implying that Q, R, and the origin P are collinear.

Proof 2. Taking P as the origin, assume
$$E = uA, \quad B = vD, \tag{1}$$
$$rA + (u-1)B = (r+u-1)C, \tag{2}$$
$$mD + (1-v)E = (m+1-v)F. \tag{3}$$
(Set up this way after trial and error, knowing that further calculations can achieve coefficient balance.) Add (2) to (3), and eliminate B and E by (1):
$$rA + (u-1)vD + mD + (1-v)uA = (r+u-1)C + (m+1-v)F. \tag{4}$$
Reorganize to get the expression for point Q:
$$(r+u-uv)A + (v-m-1)F = (r+u-1)C + (v-uv-m)D$$
$$= (r+u+v-m-uv-1)Q. \tag{5}$$
Using (1) and (3), replace A with B and F, and replace D with C and E in (5):
$$\frac{rv(m+1-v)F - (rm+um-uvm)B}{uv(1-v)}$$
$$= \frac{(rvu-rv+mr)E - mu(r+u-1)C}{uv(u-1)} = (r+u+v-m-uv-1)Q. \tag{6}$$

Note that the coefficients of the left and middle parts of (6) are equal:
$$\frac{rv(m+1-v) - (rm+um-uvm)}{uv(1-v)} = \frac{(rvu-rv+mr) - mu(r+u-1)}{uv(u-1)}$$
$$= \frac{rv - m(r+u)}{uv},$$

which means (6) gives the expression for the intersection point R of BF and CD, yielding
$$\frac{rv - m(r+u)}{uv} R = (r+u+v-m-uv-1)Q,$$
indicating that Q, R, and the origin P are collinear.

Note. Proof 1 above is a conventional method and is relatively lengthy. Proof 2 employs clever techniques, making it simpler, although its geometric significance might not be immediately apparent. If area calculations are used, a more concise and intuitive proof can be achieved.

Proof 3. Taking P as the origin, let $E = uA$ and $B = vD$.
To prove that line QR passes through point P, it suffices to show that $\frac{ERQ}{ARQ} = \frac{EP}{AP} = u$. Note that

$$ERQ = \frac{ERQ}{ECQ} \cdot \frac{ECQ}{ECD} \cdot ECD$$
$$= \frac{ER}{EC} \cdot \frac{CQ}{CD} \cdot ECD$$
$$= \frac{EBF}{EBF + BCF} \cdot \frac{ACF}{ACF + AFD} \cdot ECD,$$

$$ARQ = \frac{ARQ}{ARF} \cdot \frac{ARF}{ABF} \cdot ABF$$
$$= \frac{AQ}{AF} \cdot \frac{RF}{BF} \cdot ABF$$
$$= \frac{ACD}{ACD + DCF} \cdot \frac{CFE}{BCE + CFE} \cdot ABF.$$

Since $EBF + BCF = BCE + CFE$ and $ACF + AFD = ACD + DCF$, we have
$$\frac{ERQ}{ARQ} = \frac{ECD}{CFE} \cdot \frac{ACF}{ABF} \cdot \frac{EBF}{ACD} = \frac{DE}{FE} \cdot \frac{AC}{AB} \cdot \frac{EBF}{ACD}.$$

Using $EBF = \frac{EF}{ED}EBD$ and $ACD = \frac{AC}{AB}ABD$, we get

$$\frac{ERQ}{ARQ} = \frac{EBD}{ABD} = \frac{EP}{AP} = u.$$

Note. This approach largely avoids direct parameter calculations.

1.5 Points Multiplication by Complex Number

Using complex numbers to multiply points helps simplify problems involving angles.

Definition 6. Multiplication by imaginary unit i: If in the Cartesian coordinate system $A = (x, y)$, then define $iA = (-y, x)$, where the letter i is a reserved symbol.

Geometrically, iA is the point A being rotated 90° anticlockwise around A. Clearly, we have the following properties.

Property 14. $i(iA) = -A$.

When performing the inner product, we have the following.

Property 15. $iA \cdot A = 0$.

Property 16. For $B = (u, v)$, we get $iA \cdot B = (-y, x) \cdot (u, v) = vx - uy = -A \cdot iB$.

Definition 7. Multiplication by complex numbers: If in the Cartesian coordinate system $A = (x, y)$ and $\alpha = u + vi$ is a complex number, then define $\alpha A = uA + i(vA)$, where it's clear that $\alpha A = uA + v(iA)$.

By establishing a Cartesian coordinate system on the plane, a one-to-one correspondence can be established between points and complex numbers. Specifically, if $A = (x, y)$, then define $f(A) = x + yi$, with its inverse mapping being $p(x + yi) = (x, y) = A$. Thus, $p(f(A)) = A$ and $f(p(x + yi)) = x + yi$. After mapping points on the plane to complex numbers, it's easy to verify that the multiplication of points by complex numbers is consistent with the multiplication of complex numbers. In other words, if $A = (x, y)$ in the Cartesian coordinate system and $\alpha = u + vi$ is a complex number, then $\alpha A = p(\alpha f(A))$.

In fact, using the definition of point multiplication by complex numbers, we have

$$\alpha A = uA + v(iA) = (ux, uy) + v(-y, x) = (ux - vy, uy + vx).$$

And expressing point $A = (x, y)$ as the product of a complex number $\alpha = u + vi$ and then converting it back to coordinates, we get

$$p(\alpha f(A)) = p((u+vi)(x+yi)) = p(ux - vy + (uy + vx)i)$$
$$= (ux - vy, uy + vx).$$

Both results are equal. From this, we can deduce the following.

Property 17. Geometric interpretation of multiplication by complex numbers and points: if $A = (x, y)$ in the Cartesian coordinate system and $\alpha = u + vi$ is a complex number, let $r = |\alpha| = \sqrt{u^2 + v^2}$ and θ be the principal angle of $\alpha = u + vi$, satisfying $\alpha = r\cos\theta + ir\sin\theta$, with $0 \le \theta < 2\pi$. According to the geometric interpretation of complex number multiplication, rotating the point rA at an angle of θ radians anticlockwise around the origin yields αA.

Usually, let's denote $e^{i\theta} = \cos\theta + i\sin\theta$. Then, the point obtained by rotating point $A = (x, y)$ anticlockwise by an angle θ around the origin can be simply denoted as $e^{i\theta}A$, i.e., $e^{i\theta}A = (x\cos\theta - y\sin\theta, y\cos\theta + x\sin\theta)$. In general, it's easy to verify the following properties.

Property 18. The point obtained by rotating point A anticlockwise by an angle θ around point B is $B + e^{i\theta}(A - B)$.

Property 19. Rules of arithmetic for multiplying complex numbers and points: $(\alpha + \beta)A = \alpha A + \beta A$, $\alpha(A + B) = \alpha A + \alpha B$, $(\alpha\beta)A = (\beta\alpha)A = \beta(\alpha A)$ (α and β are complex numbers).

Property 20. ASA formula for multiplying complex numbers and points: Given a triangle with sides AB and angles $\alpha = \angle CAB$ and $\beta = \angle CBA$, when the vertices $A - B - C$ are oriented anticlockwise, the following ASA formulas hold:

$$C = A + \frac{e^{i\alpha}\sin\beta}{\sin(\alpha+\beta)}(B - A) = A + \frac{1 - e^{-2i\beta}}{1 - e^{-2i(\alpha+\beta)}}(B - A).$$

Indeed, on the one hand,

$$\frac{|AC|}{|AB|} = \frac{\sin\beta}{\sin(\alpha+\beta)} = \frac{e^{i\beta} - e^{-i\beta}}{e^{i(\alpha+\beta)} - e^{\text{anticlockwise}-i(\alpha+\beta)}} = \frac{e^{-i\alpha}(1 - e^{-2i\beta})}{1 - e^{-2i(\alpha+\beta)}}.$$

On the other hand, $C - A = |AC| \cdot e^{i\alpha}(B - A)/|AB|$, and after eliminating $\frac{|AC|}{|AB|}$ and rearranging, the result is obtained. Now, using the ASA formula, let us prove the law of sines.

Example 1 (Morley's theorem). In $\triangle ABC$, P, Q, and R are the intersections of the trisection of the angles, as shown in Figure 1.10(a). Then, $\triangle PQR$ is an equilateral triangle.

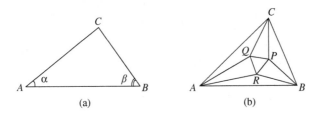

Figure 1.10

Proof 1. We want to prove the conclusion $e^{\frac{i\pi}{3}}(Q-P) = R-P$, which can be written as

$$e^{\frac{i\pi}{3}}Q - R = (e^{\frac{i\pi}{3}} - 1)P.$$

Let us denote $e^{\frac{i\pi}{3}} = \omega$, then $\omega^3 = -1$ and $\omega^2 = -\frac{1}{\omega} = \omega - 1$. The equation we want to prove becomes

$$\omega Q - R = \omega^2 P.$$

Let α, β, and γ be the angles of $\triangle ABC$, each divided by three. Without loss of generality, assume A is the origin. We can denote $e^{-2i\alpha} = u$ and $e^{-2i\beta} = v$. Using the former ASA formula, we have

$$R = \frac{1-e^{-2i\beta}}{1-e^{-2i(\alpha+\beta)}}B = \frac{1-v}{1-uv}B,$$

$$C = \frac{1-v^3}{1-u^3v^3}B,$$

$$C = \frac{1-e^{-2i(\pi-\alpha-\gamma)}}{1-e^{-2i(\pi-\gamma)}}Q = \frac{1-e^{2i\left(\frac{\pi}{3}-\beta\right)}}{1-e^{2i\left(\frac{\pi}{3}-\alpha-\beta\right)}}Q = \frac{1-\omega^2 v}{1-\omega^2 uv}Q,$$

$$P = B + \frac{1-e^{-2i\gamma}}{1-e^{-2i(\gamma+\beta)}}(C-B),$$

so

$$Q = \frac{1-\omega^2 uv}{1-\omega^2 v}C = \frac{(1-\omega^2 uv)(1-v^3)}{(1-\omega^2 v)(1-u^3 v^3)}B,$$

$$C - B = \left(\frac{1-v^3}{1-u^3 v^3} - 1\right)B = \frac{v^3(u^3-1)}{1-u^3 v^3}B,$$

$$P = \left(1 + \frac{v^3(1-\omega^{-2}u^{-1}v^{-1})(u^3-1)}{(\omega^{-2}u^{-1})(1-u^3 v^3)}\right)B = \left(1 + \frac{v^2(\omega^2 uv - 1)(u^3-1)}{(\omega^2 u - 1)(1-u^3 v^3)}\right)B.$$

Substituting on both sides of the required conclusion we want to prove, we get

$$\omega Q - R = \left(\frac{(\omega + uv)(1-v^3)}{(1-\omega^2 v)(1-u^3 v^3)} - \frac{1-v}{1-uv}\right)B,$$

$$\omega^2 P = \left(\omega^2 + \frac{\omega v^2(uv+\omega)(1-u^3)}{(\omega^2 u - 1)(1-u^3 v^3)}\right)B.$$

After simplifying, the required equality becomes

$$((\omega + uv)(1-v^3) - (1-v)(1-\omega^2 v)(1 + uv + u^2 v^2))(\omega^2 u - 1)$$
$$= (-\omega(u+\omega)(1-u^3 v^3) + \omega v^2(uv+\omega)(1-u^3))(1-\omega^2 v).$$

Expanding both sides and using the relations $\omega^3 = -1$ and $\omega^2 = \omega - 1$ to eliminate the higher powers of ω, the result on both sides simplifies to

$$\omega(u^3 v^4 - u^3 v^2 + uv^3 - u + v^3 + v^2 - v - 1)$$
$$- u^3 v^3 + u^3 v^2 + uv^4 - uv - v^2 + 1,$$

which confirms the conclusion.

Note. The above proof involves direct calculations. Using the identical method, the calculations can be simplified.

Proof 2. We use the identical method. Let $\triangle PQR$ be an equilateral triangle. If we can construct $\triangle ABC$ such that P, Q, and R are the intersections of the trisection of the interior angles, as shown in Figure 1.10(b), then the proposition is true.

Let R be the origin, then $Q = e^{\frac{i\pi}{3}}P = \omega P$. Let
$$\angle BRP = \angle CQP = \frac{\pi}{3} + \alpha,$$
$$\angle ARQ = \angle CPQ = \frac{\pi}{3} + \beta,$$
$$\angle AQR = \angle BPR = \frac{\pi}{3} + \gamma,$$
where $\alpha + \beta + \gamma = \frac{\pi}{3}$. Thus, we have
$$\angle RAQ = \alpha, \quad \angle PBR = \beta, \quad \angle PCQ = \gamma, \quad \angle ARB = \pi - \alpha - \beta.$$
To prove
$$\angle RAB = \angle QAC = \alpha,$$
$$\angle RBA = \angle PBC = \beta,$$
$$\angle PCB = \angle QCA = \gamma,$$
it suffices to prove $\angle RAB = \alpha$ and $\angle RBA = \beta$, and the other cases are similar.

Using the ASA formula, we get
$$A = R + \frac{e^{i\left(\frac{\pi}{3}+\beta\right)}\sin\left(\frac{\pi}{3}+\gamma\right)}{\sin\alpha}(Q-R) = \frac{e^{i\left(\frac{\pi}{3}+\beta\right)}\sin\left(\frac{\pi}{3}+\gamma\right)}{\sin\alpha}Q,$$
$$B = R + \frac{e^{-i\left(\frac{\pi}{3}+\alpha\right)}\sin\left(\frac{\pi}{3}+\gamma\right)}{\sin\beta}(P-R) = \frac{e^{-i\left(\frac{\pi}{3}+\alpha\right)}\sin\left(\frac{\pi}{3}+\gamma\right)}{\sin\beta}\cdot e^{-i\frac{\pi}{3}}Q$$
$$= \frac{e^{-i\left(\frac{2\pi}{3}+\alpha\right)}\sin\left(\frac{\pi}{3}+\gamma\right)}{\sin\beta}Q.$$

The desired conclusion can be written as $e^{i(\pi-\alpha-\beta)}A = \frac{\sin\beta}{\sin\alpha}B$, which is
$$\frac{e^{i(\pi-\alpha-\beta)}e^{i\left(\frac{\pi}{3}+\beta\right)}\sin\left(\frac{\pi}{3}+\gamma\right)}{\sin\alpha}Q = \frac{\sin\beta}{\sin\alpha}\cdot\frac{e^{-i\left(\frac{2\pi}{3}+\alpha\right)}\sin\left(\frac{\pi}{3}+\gamma\right)}{\sin\beta}Q.$$

After simplification, this becomes $e^{i\frac{4\pi}{3}} = e^{-i\frac{2\pi}{3}}$, which is obviously true.

1.6 Conclusion

In this chapter, we have introduced the fundamental concepts in point geometry, including point addition, scalar multiplication of points, the inner product of points, the exterior product of points, and point multiplication by complex numbers. We derived around 20 fundamental properties or formulas related to point operations, which form the basic outline of point geometry. The provided definitions, properties, formulas, and specific problem-solving examples demonstrate that point geometry not only aligns with mathematical intuition and facilitates the representation of fundamental geometric facts but also contributes to simplifying geometric reasoning.

Due to space limitations and considerations related to the national curriculum and competition scope for secondary school education, the subsequent chapters of this book will delve into the discussion of only selected properties.

Chapter 2

Computational Methods

2.1 Intersections Computing and Geometric Constructions

Can geometry problems be approached systematically, much like algebra problems?

Building upon point geometry, we present two approaches. The first approach is detailed in this chapter, where the idea is to determine individual points and subsequently consider their geometric relationships. Unlike analytical methods, this approach avoids converting points into coordinates and focuses on establishing relationships between points. This not only leads to concise point geometry calculations but also provides clearer geometric insights. The second approach is elaborated on in subsequent chapters. It involves expressing given conditions and conclusions using point geometry and establishing identities to relate them.

For a point P on line AB, its vector representation can be expressed as $\overrightarrow{OP} = t\overrightarrow{OA} + (1-t)\overrightarrow{OB}$, or, equivalently, $\overrightarrow{OP} = \frac{x\overrightarrow{OA}+y\overrightarrow{OB}}{x+y}$. In the context of linear algebra, $\overrightarrow{OP}$ is referred to as a linear combination of $\overrightarrow{OA}$ and $\overrightarrow{OB}$, where the coefficients of $\overrightarrow{OA}$ and $\overrightarrow{OB}$ sum up to 1. Because the relationship between points P, A, and B is independent of the origin's position O, for simplicity, we can write $P = tA + (1-t)B$ or $P = \frac{xA+yB}{x+y}$. This concept can be extended further. For three non-collinear points A, B, and C determining a plane, a point P on that plane can be represented as $P = tA + sB + (1-t-s)C$, or, equivalently, $P = \frac{xA+yB+zC}{x+y+z}$. Point addition and subtraction, as well as vector addition and subtraction, follow similar principles. Assuming O as the origin (often denoted as $O = 0$), the vector $\overrightarrow{AB}$ can be represented as $\overrightarrow{OB} - \overrightarrow{OA}$, which can be succinctly denoted as $B - A$. The inner product of vectors $\overrightarrow{OA} \cdot \overrightarrow{OB}$ can be simplified to $A \cdot B$

and, even more concisely, as AB (note the context to avoid confusion with line segments). In these seemingly simple abbreviations, it appears that there is nothing new, but in practice, unexpected effects are discovered.

In Figure 2.1, if we let $P = \frac{xA+yB+zC}{x+y+z}$, how can we find the points D, E, and F? Considering that D is the intersection of lines AP and BC, we have $(x+y+z)P - xA = yB + zC = (y+z)D$, which immediately yields $D = \frac{yB+zC}{y+z}$. Similarly, $E = \frac{xA+zC}{x+z}$ and $F = \frac{xA+yB}{x+y}$. Let us compute point Q using three methods.

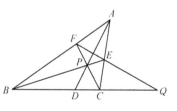

Figure 2.1

Method 1. Assume $Q = tF + (1-t)E = sB + (1-s)C$, i.e.,

$$t\frac{xA+yB}{x+y} + (1-t)\frac{xA+zC}{x+z} = sB + (1-s)C.$$

Solve the system of equations in terms of the coefficients of A, B, and C:

$$\begin{cases} t\dfrac{x}{x+y} + (1-t)\dfrac{x}{x+z} = 0, \\ t\dfrac{y}{x+y} = s, \\ (1-t)\dfrac{z}{x+z} = 1-s. \end{cases}$$

This yields

$$\begin{cases} t = \dfrac{x+y}{y-z}, \\ s = \dfrac{y}{y-z}. \end{cases}$$

Thus,

$$Q = sB + (1-s)C = \frac{yB - zC}{y - z}.$$

Method 2. Assume $Q = tF + (1-t)E$, i.e.,

$$t\frac{xA+yB}{x+y} + (1-t)\frac{xA+zC}{x+z} = x\left(\frac{t}{x+y} + \frac{1-t}{x+z}\right)A$$
$$+ \frac{ty}{x+y}B + \frac{(1-t)z}{x+z}C.$$

Since Q lies on line BC, $x\left(\frac{t}{x+y} + \frac{1-t}{x+z}\right) = 0$, which gives $t = \frac{x+y}{y-z}$. Thus,

$$Q = tF + (1-t)E = \frac{yB - zC}{y - z}.$$

Method 3. As Q is the intersection of lines EF and BC, subtracting the equations $(x+z)E = xA + zC$ and $(x+y)F = xA + yB$ yields $(x+z)E - (x+y)F = zC - yB = (z-y)Q$. Thus, $Q = \frac{zC - yB}{z - y}$.

E and F were originally represented in terms of A, B, and C, while in Method 3, A was easily eliminated to obtain the intersection point Q of BC and EF. This technique is crucial. Method 2 only requires setting a parameter t and solving a single equation, which is evidently simpler than method 1. However, method 1 is a general approach for finding intersection points and also needs to be mastered.

With the coordinates of the points established above, many tasks can be easily accomplished.

Proof of Ceva's theorem:

$$\frac{|BD|}{|DC|} \cdot \frac{|CE|}{|EA|} \cdot \frac{|AF|}{|FB|} = \frac{z}{y} \cdot \frac{x}{z} \cdot \frac{y}{x} = 1.$$

Proof of Menelaus' theorem:

$$\frac{|AF|}{|FB|} \cdot \frac{|BQ|}{|QC|} \cdot \frac{|CE|}{|EA|} = \frac{y}{x} \cdot \frac{z}{y} \cdot \frac{x}{z} = 1.$$

Proof of the projection theorem property:

$$\frac{|BD|}{|DC|} = \frac{|BQ|}{|CQ|} = \frac{z}{y}.$$

$P = \frac{xA + yB + zC}{x+y+z}$, taking the exterior product of P with AB, we obtain $PAB = \frac{zCAB}{x+y+z}$. Similarly, $PBC = \frac{xABC}{x+y+z}$ and $PCA = \frac{yBCA}{x+y+z}$. It is evident that the coefficients $x : y : z = S_{\triangle PBC} : S_{\triangle PCA} : S_{\triangle PAB}$, and this property can be utilized to determine properties such as the triangle's incentre, circumcentre, centroid, orthocentre, and excentre, as well as other coincidences of triangles.

In Figure 2.1, the following is proven: $S_{\triangle BPC}\overrightarrow{PA} + S_{\triangle CPA}\overrightarrow{PB} + S_{\triangle APB}\overrightarrow{PC} = \vec{0}$. For this purpose, it suffices to note that

$$x(A - P) + y(B - P) + z(C - P) = xA + yB + zC - (x+y+z)P = 0,$$

which implies $S_{\triangle BPC}(A - P) + S_{\triangle CPA}(B - P) + S_{\triangle APB}(C - P) = 0$, resulting in

$$S_{\triangle BPC}\overrightarrow{PA} + S_{\triangle CPA}\overrightarrow{PB} + S_{\triangle APB}\overrightarrow{PC} = \vec{0}.$$

Another conclusion is

$$P = \frac{S_{\triangle BPC}A + S_{\triangle CPA}B + S_{\triangle APB}C}{S_{\triangle BPC} + S_{\triangle CPA} + S_{\triangle APB}}.$$

It is also possible to prove using the vector method and the area method:

$$\overrightarrow{AP} = \frac{AP}{AD}\overrightarrow{AD} = \frac{AP}{AD}\frac{DC}{BC}\overrightarrow{AB} + \frac{AP}{AD}\frac{BD}{BC}\overrightarrow{AC}$$

$$= \frac{S_{\triangle CPA}}{S_{\triangle ACD}}\frac{S_{\triangle ACD}}{S_{\triangle ABC}}\overrightarrow{AB} + \frac{S_{\triangle APB}}{S_{\triangle ABD}}\frac{S_{\triangle ABD}}{S_{\triangle ABC}}\overrightarrow{AC}$$

$$= \frac{S_{\triangle CPA}}{S_{\triangle ABC}}\overrightarrow{AB} + \frac{S_{\triangle APB}}{S_{\triangle ABC}}\overrightarrow{AC}$$

$$= \frac{S_{\triangle CPA}}{S_{\triangle ABC}}(\overrightarrow{PB} - \overrightarrow{PA}) + \frac{S_{\triangle APB}}{S_{\triangle ABC}}(\overrightarrow{PC} - \overrightarrow{PA}).$$

Hence,

$$S_{\triangle BPC}\overrightarrow{PA} + S_{\triangle CPA}\overrightarrow{PB} + S_{\triangle APB}\overrightarrow{PC} = \vec{0}.$$

The following properties are easily deduced:

(1) If point P is the centroid, then

$$S_{\triangle BPC} = S_{\triangle CPA} = S_{\triangle APB}, \overrightarrow{PA} + \overrightarrow{PB} + \overrightarrow{PC} = \vec{0},$$

and $P = \frac{A+B+C}{3}$.

(2) If point P is the incentre, then $a\overrightarrow{PA} + b\overrightarrow{PB} + c\overrightarrow{PC} = \vec{0}$, or alternatively written as

$$(\sin A)\overrightarrow{PA} + (\sin B)\overrightarrow{PB} + (\sin C)\overrightarrow{PC} = \vec{0},$$

and $P = \frac{aA+bB+cC}{a+b+c}$.

(3) If point P is the excentre corresponding to point A, then
$$-a\overrightarrow{PA} + b\overrightarrow{PB} + c\overrightarrow{PC} = \vec{0},$$
and $P = \frac{-aA+bB+cC}{-a+b+c}$.

(4) If point P is the circumcentre, then
$$\overrightarrow{PA}\sin 2A + \overrightarrow{PB}\sin 2B + \overrightarrow{PC}\sin 2C = \vec{0},$$
and
$$P = \frac{\sin 2A \cdot A + \sin 2B \cdot B + \sin 2C \cdot C}{\sin 2A + \sin 2B + \sin 2C}.$$

(5) If point P is the orthocentre of an oblique triangle, then
$$\overrightarrow{PA}\tan A + \overrightarrow{PB}\tan B + \overrightarrow{PC}\tan C = \vec{0},$$
and
$$P = \frac{\tan A \cdot A + \tan B \cdot B + \tan C \cdot C}{\tan A + \tan B + \tan C}.$$

For any point K on line AD with the form $K = kA + (1-k)\frac{yB+zC}{y+z}$, regardless of how k varies, the ratio of coefficients of B and C remains constant as $y:z$. This relates to the co-side theorem: $\frac{|BD|}{|DC|} = \frac{|S_{\triangle AKB}|}{|S_{\triangle AKC}|}$.

Geometric research spanning thousands of years has yielded rich results in the field of compass and straightedge construction. Researchers have discovered that many seemingly complex geometric figures can be constructed from a small set of basic geometric constructions. Conversely, complex figures can also be broken down into simpler components. The most fundamental geometric constructions include dividing a line segment into a given ratio, constructing a perpendicular line through a given point, and constructing a line segment equal to a known length. Additionally, for problem-solving convenience, it is possible to expand upon the basic geometric constructions to establish more advanced constructions. For example, constructing a parallelogram given three points is achieved by composing two parallel line constructions and one intersection point construction. The following table presents some basic constructions, which can be further expanded upon as needed.

Point Geometry Expression	Geometric Meaning
$(A-B)^2 = (C-D)^2$	Equal segments: $AB = CD$
$(A-B)^2 = (A-C)^2$, or $\left(A - \frac{B+C}{2}\right)(B-C) = 0$	Equal segments: $AB = AC$
$C = tA + (1-t)B$	Point C on line AB, $\overrightarrow{BC} = t\overrightarrow{BA}$
Implicit: $2C - A - B = 0$, Explicit: $C = \frac{A+B}{2}$	C is the midpoint of AB
$(A-B)(C-D) = 0$	$AB \perp CD$
$C - D = t(A-B)$	$CD \parallel AB$
$OA^2 = OB^2$ (Equivalent to $\left(O - \frac{A+B}{2}\right)(A-B) = 0$), $OB^2 = OC^2$, $OA^2 = OC^2$. At least two of these equivalent expressions	O is the circumcentre of $\triangle ABC$
$(A-H)(B-C) = 0$, $(B-H)(C-A) = 0$, $(C-H)(A-B) = 0$. Two of these three equations are used.	H is the orthocentre of $\triangle ABC$
$H = A + B + C$	If the circumcentre O of $\triangle ABC$ is taken as the origin, and H is the orthocentre of $\triangle ABC$.
$K = \frac{A+B+C}{2}$	If the circumcentre O of $\triangle ABC$ is taken as the origin, and K is the nine-point centre of $\triangle ABC$.
$G = \frac{A+B+C}{3}$	G is the centroid of $\triangle ABC$
$B - A = C - D$ and its equivalent $A + C = B + D$	Parallelogram $ABCD$
$(A-D)^2 - (B-D)(D-C) = 0$ or $(A-B)^2 - (B-D)(B-C) = 0$	Right-angled triangle projection theorem: In $\triangle ABC$, $AB \perp AC$, $AD \perp BC$
$(P-A)(P-B) - (P-C)(P-D) = 0$ or $(P-A)(P-B) = (P-O)^2 - R^2$	Circular Power Theorem: In cyclic quadrilateral $ABCD$, line AB intersects CD at P, where R is the radius

Note: Please remember the above properties, as they are frequently used and help simplify problem-solving.

2.2 Point-Line Positions

Example 1. Given points P and Q inside $\triangle ABC$ and
$$\overrightarrow{PA} + 2\overrightarrow{PB} + 3\overrightarrow{PC} = 2\overrightarrow{QA} + 3\overrightarrow{QB} + 5\overrightarrow{QC} = \vec{0},$$
find $\frac{|\overrightarrow{PQ}|}{|\overrightarrow{AB}|}$. (2018 National Secondary School Mathematics Competition Liaoning Province Preliminary)

Solution. $P = \frac{A+2B+3C}{6}$, $Q = \frac{2A+3B+5C}{10}$, $P - Q = \frac{1}{30}(-A+B)$, $\frac{|\overrightarrow{PQ}|}{|\overrightarrow{AB}|} = \frac{1}{30}$.

Note. It can be derived from the proof that $PQ \parallel AB$.

Example 2. In $\triangle ABC$, as shown in Figure 2.2, H is the orthocentre, O is the circumcentre, A_1 is the midpoint of BC, S is symmetric to H about point A, and L is symmetric to A about A_1. Prove that S is symmetric to L about O.

Proof. Let $O = 0$, $H = A + B + C$, $L = B + C - A$, and $S = 2A - (A + B + C)$, which gives $L + S = 0$.

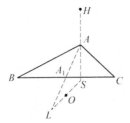
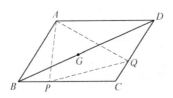

Figure 2.2 Figure 2.3

Example 3. In parallelogram $ABCD$, as shown in Figure 2.3, points P and Q lie on BC and CD, respectively, and $\frac{BP}{PC} = \frac{CQ}{QD}$. Prove that the centroid of $\triangle APQ$ lies on BD.

Proof. Let $P = tB + (1-t)C$ and $Q = tC + (1-t)D$:
$$\frac{(B+D-C)+P+Q}{3} = \frac{1}{3}(B + 2D + tB - tD).$$

Example 4. In parallelogram $ABCD$, as shown in Figure 2.4, point M is symmetric to point P about A, point N is symmetric to point M about D, and point Q is symmetric to point N about C. Prove that point P is symmetric to point Q about B.

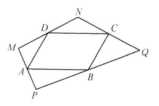

Figure 2.4

Translation. Given that $M + P = 2A$, $N + M = 2D$, $Q + N = 2C$, and $A + C = B + D$, prove that $P + Q = 2B$.

Proof:

$$P + Q = (2A - M) + (2C - N)$$
$$= (2A + 2C) - (M + N)$$
$$= 2A + 2C - 2D = 2B.$$

Note. This problem can be seen as the converse proposition of "a quadrilateral formed by connecting the midpoints of consecutive sides of a quadrilateral is a parallelogram". The difficulty level is a bit higher, and it is important to note that P does not necessarily lie on the plane of $ABCD$.

Example 5. In $\triangle ABC$, as shown in Figure 2.5, point A_1 is chosen on ray BA such that $BA_1 = BC$, and point A_2 is chosen on ray CA such that $CA_2 = BC$. Similarly, points B_1, B_2, C_1, and C_2 are defined. Prove that $A_1A_2 \parallel B_1B_2 \parallel C_1C_2$. (2008 Sharygin Geometry Olympiad)

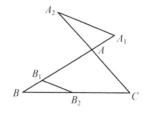

Figure 2.5

Proof. Let $A = 0$, then $A_1 = -\frac{a-c}{c}B$ and $A_2 = -\frac{a-b}{b}C$. Also, $B_1 = \frac{b}{c}B$ and $B_2 = \frac{bB+(a-b)C}{a}$.

$$A_2 - A_1 = \frac{b(a-c)B - (a-b)cC}{bc},$$

$$B_1 - B_2 = \frac{b(a-c)B - (a-b)cC}{ac}.$$

Therefore, $A_1A_2 \parallel B_1B_2$. Similarly, we can prove $B_1B_2 \parallel C_1C_2$.

Note. From the proof, we obtain $\frac{A_1A_2}{B_1B_2} = \frac{a}{b}$.

Example 6. Prove that the side lengths of a triangle form an arithmetic sequence if and only if the line connecting its centroid and incentre is parallel to one of the sides.

Proof.
$$I - G = \frac{aA + bB + cC}{a + b + c} - \frac{A + B + C}{3}$$
$$= \frac{(2a - b - c)A + (2b - c - a)B + (2c - a - b)C}{3(a + b + c)}.$$

If $2a - b - c = 0$, then $I - G = \frac{b-c}{2(a+b+c)}(B - C)$. Conversely, if $IG \parallel BC$, then $2a - b - c = 0$ and $2b - c - a = -(2c - a - b)$, which leads to $2a - b - c = 0$. Similar arguments apply to the other two cases.

Example 7. In $\triangle ABC$, as shown in Figure 2.6, O is the centre of the equilateral $\triangle ABC$. Point M lies on side BC, and K and L are the projections of M onto AB and AC, respectively. Prove that line OM bisects segment KL. (2006 Balkan Mathematical Olympiad (JBMO) Preliminary Problems)

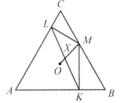

Figure 2.6

Proof. Let $AB = 1$ and $CM = t$, then $O = \frac{A+B+C}{3}$:

$$M = tB + (1 - t)C,$$

$$L = \frac{t}{2}A + \left(1 - \frac{t}{2}\right)C,$$

$$K = \frac{1-t}{2}A + \left(1 - \frac{1-t}{2}\right)B;$$

$$\frac{L + K}{2} = \frac{\frac{t}{2}A + \left(1 - \frac{t}{2}\right)C + \frac{1-t}{2}A + \left(1 - \frac{1-t}{2}\right)B}{2}$$
$$= \frac{1}{4}A + \frac{1+t}{4}B + \frac{2-t}{4}C$$
$$= \frac{3}{4}\cdot\frac{A + B + C}{3} + \frac{1}{4}[tB + (1 - t)C]$$
$$= \frac{3O + M}{4}.$$

Example 8. In $\triangle ABC$, as shown in Figure 2.7, let AM and BN be the two angle bisectors of $\triangle ABC$. Point P lies on segment MN, and the distances from P to BC, CA, and AB are PD, PE, and PF, respectively. Prove that $PD+PE = PF$.

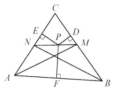

Figure 2.7

Proof. Let $M = \frac{bB+cC}{b+c}$ and $N = \frac{aA+cC}{a+c}$. Then,

$$P = tM + (1-t)N = \frac{a(b+c-bt-ct)A + bt(a+c)B + c(b+c+at-bt)C}{(a+c)(b+c)}.$$

To prove $PD + PE = PF$, we only need to prove that $\frac{\triangle PBC}{a} + \frac{\triangle PCA}{b} = \frac{\triangle PAB}{c}$. Simplifying the equation,

$$(b+c-bt-ct) + t(a+c) = b+c+at-bt.$$

Note. In point geometry, since the concept of line segments does not exist, proving that two line segments are equal often involves proving that the squares of the line segments are equal. To prove that the sum of two line segments is equal to another segment, unless the positions of these three line segments are very special, it is generally not easy. The key to this problem lies in understanding the geometric meaning of the coefficients of point P, where the ratios of the coefficients represent the ratios of the triangles $\triangle PBC$, $\triangle PCA$, and $\triangle PAB$.

Example 9. In $\triangle ABC$, as shown in Figure 2.8, let I be the incentre of triangle ABC, and D, E, and F be the midpoints of sides BC, CA, and AB, respectively. Prove that line DI bisects the perimeter of $\triangle DEF$. (2018 Jiangxi Provincial Preliminary Contest of National High School Mathematics League)

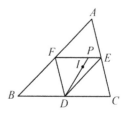

Figure 2.8

Proof 1. Let $P = t\frac{B+C}{2} + (1-t)\frac{aA+bB+cC}{a+b+c} = s\frac{A+B}{2} + (1-s)\frac{A+C}{2}$. In other words,

$$\left(\frac{1}{2}(-1+s) - \frac{s}{2} + \frac{a(1-t)}{a+b+c}\right)A + \left(-\frac{s}{2} + \frac{b(1-t)}{a+b+c} + \frac{t}{2}\right)B$$
$$+ \left(\frac{1}{2}(-1+s) + \frac{c(1-t)}{a+b+c} + \frac{t}{2}\right)C = 0.$$

Solving the equation

$$\frac{1}{2}(-1+s) - \frac{s}{2} + \frac{a(1-t)}{a+b+c} = -\frac{s}{2} + \frac{b(1-t)}{a+b+c} + \frac{t}{2}$$

$$= \frac{1}{2}(-1+s) + \frac{c(1-t)}{a+b+c} + \frac{t}{2} = 0,$$

we get $t = \frac{a-b-c}{2a}$ and $s = \frac{a+b-c}{2a}$. Therefore,

$$|DF| + |FP| = \frac{b}{2} + \frac{a}{2}\left(1 - \frac{a+b-c}{2a}\right) = \frac{1}{4}(a+b+c) = \frac{1}{2}\frac{a+b+c}{2}.$$

Proof 2. We have

$$I = \frac{aA + bB + cC}{a+b+c} = \frac{a(E+F-D) + b(-E+F+D) + c(E-F+D)}{a+b+c}$$

$$= \frac{(-a+b+c)D + (a-b+c)E + (a+b-c)F}{a+b+c},$$

so

$$P = \frac{(a+b+c)I - (-a+b+c)D}{2a} = \frac{(a-b+c)E + (a+b-c)F}{2a}.$$

The following steps are similar to the above proof.

Example 10. In $\triangle ABC$, as shown in Figure 2.9, let E and F be points on sides AC and AB, respectively. Lines BE and CF intersect at point D. Lines AD and EF intersect at point G. Lines parallel to BC passing through point D intersect AB, BG, CG, and AC at points H, K, N, and M, respectively. Prove that $2KN = HM$. (Problems for Solution 2066 in Mathematical Bulletin)

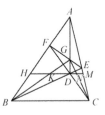

Figure 2.9

Proof. Let $D = \frac{xA+yB+zC}{x+y+z}$. Then, $E = \frac{xA+zC}{x+z}$ and $F = \frac{xA+yB}{x+y}$.

Let $G = tE + (1-t)F = \left[\frac{(1-t)x}{x+y} + \frac{tx}{x+z}\right]A + \frac{(1-t)y}{x+y}B + \frac{tz}{x+z}C$. Since G lies on AD, we have $\frac{\frac{(1-t)y}{x+y}}{\frac{tz}{x+z}} = \frac{y}{z}$, which gives $t = \frac{x+z}{2x+y+z}$. Thus, $G = \frac{2xA+yB+zC}{2x+y+z}$.

For point H, we have $H = D + k(B-C) = \frac{x}{x+y+z}A + \left(k + \frac{y}{x+y+z}\right)B + \left(\frac{z}{x+y+z} - k\right)C$. Since H lies on AB, we find $k = \frac{z}{x+y+z}$, so $H = \frac{xA+(y+z)B}{x+y+z}$.

For point K, since K lies on BG, we can set $K = \frac{2xA+rB+zC}{2x+r+z}$. By equating the coefficient of A in $H - K$, we find $r = 2y + z$, leading to $K = \frac{2xA+(2y+z)B+zC}{2(x+y+z)}$.

By symmetry, we have $M = \frac{xA+(y+z)C}{x+y+z}$ and $N = \frac{2xA+yB+(y+2z)C}{2(x+y+z)}$. It is easy to verify that $2(K - N) = H - M$.

Example 11. In $\triangle ABC$, as shown in Figure 2.10, a line from point A intersects the extension of side BC at D. Choose any point P on line AD. Line BP intersects AC at E, and line CP intersects the extension of BA at F. Line FG, parallel to BC, is drawn through F, intersecting the extension of DE at G. Prove that line FG bisects segment DA.

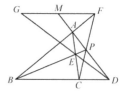

Figure 2.10

Proof. Let

$$P = \frac{xA + yB + zC}{x + y + z}, \quad D = \frac{yB + zC}{y + z},$$

$$E = \frac{xA + zC}{x + z}, \quad F = \frac{xA + yB}{x + y},$$

$$M = \frac{xA + yD}{x + y} = \frac{xA + y\frac{yB+zC}{y+z}}{x + y}.$$

Calculate $2M - F = tD + (1-t)E$. When $t = \frac{y-z}{x+y}$, the equality holds.

Note. Use the property of similar triangles to derive $M = \frac{xA+yD}{x+y}$ from $F = \frac{xA+yB}{x+y}$.

Example 12. As shown in Figure 2.11, AD, BE, and CF are the altitudes of $\triangle ABC$, and G, I, J, N, M, and L are the midpoints of AB, BC, CA, AD, BE, and CF, respectively. Prove that lines NI, GL, and MJ are concurrent.

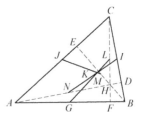

Figure 2.11

Proof. We have $2G = A + B$, $F = \frac{(-b^2+a^2+c^2)A+(b^2-a^2+c^2)B}{2c^2}$, $2L = C + F$, $2I = C + B$, and $D = \frac{(-c^2+b^2+a^2)B+(c^2-b^2+a^2)C}{2a^2}$, $2N = A + D$.

Let NI intersect LG at point K. $K = mN + (1-m)I$ and $K = nL + (1-n)G$. This yields the following equations:

$$m\left[A + \frac{(-c^2+b^2+a^2)B + (c^2-b^2+a^2)C}{2a^2}\right] + (1-m)(C+B)$$

$$= n\left[C + \frac{(-b^2+a^2+c^2)A + (b^2-a^2+c^2)B}{2c^2}\right] + (1-n)(A+B)$$

Solving the system of equations,

$$\begin{cases} m = n\dfrac{-b^2+a^2+c^2}{2c^2} + (1-n), \\ m\dfrac{c^2-b^2+a^2}{2a^2} + (1-m) = n, \end{cases}$$

we find that $m = \frac{2a^2}{a^2+b^2+c^2}$.

Therefore,

$$K = \frac{2a^2}{a^2+b^2+c^2}\left(A + \frac{(-c^2+b^2+a^2)B + (c^2-b^2+a^2)C}{2a^2}\right)$$

$$+ \left(1 - \frac{2a^2}{a^2+b^2+c^2}\right)\frac{C+B}{2} = \frac{a^2A + b^2B + c^2C}{a^2+b^2+c^2}.$$

Due to symmetry, it can be inferred that point K also lies on line MJ.

Generalization. In $\triangle ABC$, let AD, BE, and CF intersect at H. Let G, I, J, N, M, and L be the midpoints of AB, BC, CA, AD, BE, and CF, respectively. Prove that NI, GL, and MJ are concurrent.

Proof. Let $H = \frac{xA+yB+zC}{x+y+z}$:

$$\frac{x(y+z)A + y(x+z)B + z(x+y)C}{2(xy+xz+yz)}$$

$$= \frac{x(y+z)}{xy+xz+yz} \cdot \frac{A + \frac{yB+zC}{y+z}}{2} + \left(1 - \frac{x(y+z)}{xy+xz+yz}\right)\frac{B+C}{2}$$

$$= \frac{y(x+z)}{xy+xz+yz} \cdot \frac{B + \frac{xA+zC}{x+z}}{2} + \left(1 - \frac{y(x+z)}{xy+xz+yz}\right)\frac{A+C}{2}$$

$$= \frac{z(x+y)}{xy+xz+yz} \cdot \frac{C + \frac{xA+yB}{x+z}}{2} + \left(1 - \frac{z(x+y)}{xy+xz+yz}\right)\frac{A+B}{2}.$$

In particular, when $x = \tan A$, $y = \tan B$, and $z = \tan C$, then

$$\frac{x(y+z)A + y(x+z)B + z(x+y)C}{2(xy+xz+yz)} = \frac{a^2 A + b^2 B + c^2 C}{a^2 + b^2 + c^2}.$$

This concludes the proof.

Exercise 2.2

1. In Figure 2.12, there is a point C on the line segment AB and a point D outside AB. Points E, F, and G are the midpoints of CD, BD, and AB, respectively. Point H is the midpoint of EG, and FH intersects AB at point I. Prove that I is the midpoint of AC.

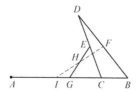

Figure 2.12

2. In Figure 2.13, AD is an altitude of the Rt$\triangle ABC$ on the hypotenuse BC. P is the midpoint of AD. Connect BP, and extend it to intersect AC at E. Given that $AC : AB = k$, find $AE : EC$. (1999 Shandong Province Competition)

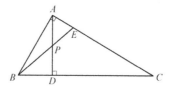

Figure 2.13

3. In Figure 2.14, in trapezium $ABCD$, AB is parallel to CD, and $AB = 3CD$. E is the midpoint of diagonal AC. Line BE intersects AD at F. Find the value of $AF : FD$. (1996 Huanggang Area Competition in China)

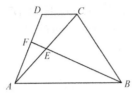

Figure 2.14

4. A median BK and an angle bisector CL of $\triangle ABC$ intersect at point P. Prove that $\frac{PC}{PL} - \frac{AC}{BC} = 1$.
5. In Figure 2.15, I is the incentre of $\triangle ABC$, D is the point where the incircle touches side BC, and the line BI intersects AC at point E. Prove that $AB \perp AC$ if and only if $\frac{BI}{BD} = \frac{BE}{AB}$.

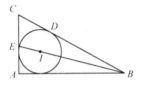

Figure 2.15

6. In Figure 2.16, quadrilateral $ABCD$ has all sides equal, and $\angle B = 60°$. A certain line passing through point D intersects the extensions of BA and BC at E and F, respectively, but does not intersect quadrilateral $ABCD$ (except at D). Point M is the intersection of lines CE and AF. Prove that $CA^2 = CM \cdot CE$.

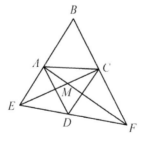

Figure 2.16

7. In Figure 2.17, for any point O inside $\triangle ABC$, lines $MN \parallel BC$, $PQ \parallel BA$, and $RS \parallel AC$ are drawn through O. Prove that $\frac{MN}{BC} + \frac{SR}{CA} + \frac{QP}{AB} = 2$.

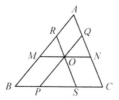

Figure 2.17

8. In Figure 2.18, prove that the Euler lines of four triangles in an orthocentric quadrilateral intersect are concurrent. (In geometry, an orthocentric quadrilateral is a quadrilateral in which one point is the orthocentre of each of the four triangles formed by three of the vertices. If H is the orthocentre of $\triangle ABC$, then A, B, and C are the orthocentres of $\triangle HBC$, $\triangle HCA$, and $\triangle HAB$, respectively.)

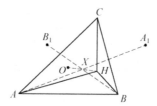

Figure 2.18

9. In Figure 2.19, in $\triangle ABC$, with $BD = CE$, CD intersects BE at G, and F is the midpoint of BC. If F is the midpoint of GM, prove that AM is the angle bisector of $\angle BAC$.

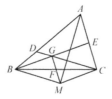

Figure 2.19

2.3 Perpendicular and Equal Segments

Example 1. As shown in Figure 2.20, in quadrilateral $ABCD$, where $AD \neq BC$, and $\frac{AE}{EB} = \frac{FA}{FB} = \frac{DG}{GC} = \frac{HD}{HC} = \frac{AD}{BC}$, prove that EG is perpendicular to FH.

Proof. Let $\left|\frac{AD}{BC}\right| = t$, then $A - E = t(E - B)$, which implies

$$E = \frac{A + tB}{1 + t}.$$

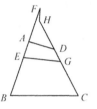

Figure 2.20

Similarly, $F = \frac{A-tB}{1-t}$, $G = \frac{D+tC}{1+t}$, and $H = \frac{D-tC}{1-t}$, $(E-G)(F-H) = \frac{(A-D)^2 - t^2(B-C)^2}{(1-t)(1+t)} = 0$.

Example 2. As shown in Figure 2.21, in $\triangle ABC$, where $AB = BC$, let D be a point on the extension of AB and E be a point on the extension of BC such that $CE = AD$. Extend AC to meet DE at F. Draw a line FG parallel to BE that intersects CD at G, and draw a line FH parallel to AD that intersects AE at H. Prove that $FG = FH$ and $AF \perp GH$. (Problems for Solution 2382 in Mathematical Bulletin)

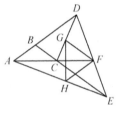

Figure 2.21

Proof. Let $B = 0$, $D = -mA$, $E = nC$, and $F = tD + (1-t)E = -mtA + n(1-t)C$. Solving $-mt + n(1-t) = 1$ yields $t = \frac{n-1}{m+n}$, and $F = \frac{n-1}{m+n}D + \left(1 - \frac{n-1}{m+n}\right)E$. According to the properties of parallel lines, we have

$$H = \frac{n-1}{m+n}A + \left(1 - \frac{n-1}{m+n}\right)E = \frac{(-1+n)A + (1+m)nC}{m+n},$$

$$G = \frac{n-1}{m+n}D + \left(1 - \frac{n-1}{m+n}\right)C = \frac{m(1-n)A + (m+1)C}{m+n},$$

$$\frac{G+H}{2} = \frac{(-1+m+n-mn)A + (1+m+n+mn)C}{2(m+n)},$$

and the coefficients sum to $\frac{(-1+m+n-mn)+(1+m+n+mn)}{2(m+n)} = 1$. Therefore, $\frac{G+H}{2}$ lies on line AC. Moreover, $(A-C)(G-H) = \frac{(1+m)(1-n)(A^2-C^2)}{m+n} = 0$, so $FG = FH$, and $AF \perp GH$.

Note. When dealing with isosceles triangles, it is often convenient to set one vertex as the origin for calculations. Utilizing the properties of isosceles triangles where the three medians coincide can simplify the calculations. For example, computing $(A - C)(G - H)$ is simpler than computing $\left(F - \frac{G+H}{2}\right)(G - H)$. It is also worth noting that the condition "$CE = AD$" in the problem is redundant. Additionally, a new proposition is derived: in $\triangle ABC$, where D is a point on the extension of AB and E is a point on the extension of BC, extending AC to intersect DE at F, and drawing lines FG parallel to BE and FH parallel to AD, prove that $AB = BC \iff AF \perp GH$.

Example 3. As shown in Figure 2.22, in $\triangle ABC$, where $\angle C = 90°$, extend AC to A_1 and extend BC to B_1 so that $AA_1 = BB_1 = AB$. Let I be the incentre of $\triangle ABC$ and M be the midpoint of AB. Prove that $MI \perp A_1 B_1$. (Problems for Solution 1114 in Mathematical Bulletin)

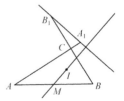

Figure 2.22

Proof. Let $C = 0$, $A_1 = -\frac{c-b}{b}A$, $B_1 = -\frac{c-a}{a}B$, and $I = \frac{aA+bB+cC}{a+b+c}$. Then,

$$\left(I - \frac{A+B}{2}\right)(A_1 - B_1) = \frac{a(a-b-c)(b-c)A^2 + b(a-c)(a-b+c)B^2}{2ab(a+b+c)}$$

$$= \frac{ab^2(a-b-c)(b-c) + ba^2(a-c)(a-b+c)}{2ab(a+b+c)}$$

$$= \frac{(a-b)(a^2+b^2-c^2)}{2(a+b+c)} = 0.$$

Note. For simplicity, we used $\angle C = 90°$ during the calculations, eliminating the term AB. If you prefer, you can keep this term until the end, resulting in a new proposition:

$$\left(I - \frac{A+B}{2}\right)(A_1 - B_1)$$

$$= \frac{a(a-b-c)(b-c)A^2 + b(a-c)(a-b+c)B^2 - (a-b)(2ab-ac-bc-c^2)AB\frac{a^2+b^2-c^2}{2}}{2ab(a+b+c)}$$

$$= \frac{a(a-b-c)(b-c)b^2 + b(a-c)(a-b+c)a^2 - (a-b)c(a^2+b^2-c^2)}{4ab}.$$

New Proposition. In $\triangle ABC$, as described in the figure, extend AC to A_1 and extend BC to B_1 such that $AA_1 = BB_1 = AB$. Let I be the incentre of $\triangle ABC$ and M be the midpoint of AB. Prove that $MI \perp A_1B_1$ if and only if $\angle C = 90°$ or $CA = CB$.

Example 4. As shown in Figure 2.23, in $\triangle ABC$, the inscribed circle touches the sides BC, CA, and AB at D, E, and F, respectively. Point G lies on segment AB, and $AF = GB$. Prove that $AB \perp AC$ if and only if $DF \perp EG$.

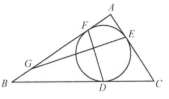

Figure 2.23

Proof. Let $A = 0$:

$$\left(\frac{\frac{a-b+c}{2}A + \frac{-a+b+c}{2}B}{c} - \frac{\frac{a-b+c}{2}C + \frac{a+b-c}{2}B}{a}\right)$$

$$\times \left(\frac{\frac{a+b-c}{2}A + \frac{-a+b+c}{2}C}{b} - \frac{\frac{-a+b+c}{2}A + \frac{a-b+c}{2}B}{c}\right)$$

$$= \frac{(a-c)(a-b+c)^2}{4ac^2}B^2$$

$$+ \frac{(a^3 - a^2b - ab^2 + b^3 - a^2c + 4abc - 3b^2c - ac^2 + bc^2 + c^3)}{4abc}BC$$

$$+ \frac{(a-b-c)(a-b+c)}{4ab}C^2$$

$$= \frac{(a-c)(a-b+c)^2}{4ac^2}c^2$$

$$+ \frac{(a^3 - a^2b - ab^2 + b^3 - a^2c + 4abc - 3b^2c - ac^2 + bc^2 + c^3)}{4abc} \cdot \frac{b^2 + c^2 - a^2}{2}$$

$$+ \frac{(a-b-c)(a-b+c)}{4ab}b^2$$

$$= \frac{(-a+b+c)(a+b-c)(a-b+c)(a^2 - b^2 - c^2)}{8abc},$$

so $AB \perp AC \Leftrightarrow DF \perp EG$.

Example 5. Let N, I, and G be the nine-point centre, incentre, and centroid of $\triangle ABC$, respectively. Prove that $NG \perp AI \Leftrightarrow \angle A = 60°$. (The nine-point centre of a triangle is the midpoint of the line segment joining the orthocentre and the circumcentre.)

Proof 1. Let $O = 0$, $N = \frac{A+B+C}{2}$, $I = \frac{aA+bB+cC}{a+b+c}$, and $G = \frac{A+B+C}{3}$:

$$(N-G)(A-I)$$
$$= \frac{(b+c)A^2 - bB^2 - cC^2 + cAB + bAC - (b+c)BC}{6(a+b+c)}$$
$$= \frac{(b+c)R^2 - bR^2 - cR^2 + c\frac{2R^2-c^2}{2} + b\frac{2R^2-b^2}{2} - (b+c)\frac{2R^2-a^2}{2}}{6(a+b+c)}$$
$$= \frac{(b+c)(a^2 - b^2 + bc - c^2)}{12(a+b+c)} = 0,$$

so $a^2 - b^2 + bc - c^2 = 0 \Leftrightarrow \angle A = 60°$.

Proof 2. Let $A = 0$, $N = \frac{H+O}{2}$, $I = \frac{aA+bB+cC}{a+b+c}$, and $G = \frac{A+B+C}{3}$:

$$(N-G)(A-I)$$
$$= \frac{2bB^2 + 2cC^2 + 2(b+c)BC - 3bBH - 3cCH - 3bBO - 3cCO}{6(a+b+c)}$$
$$= \frac{2bc^2 + 2cb^2 + 2(b+c)\frac{b^2+c^2-a^2}{2} - 3b\frac{b^2+c^2-a^2}{2} - 3c\frac{b^2+c^2-a^2}{2} - 3b\frac{c^2}{2} - 3c\frac{b^2}{2}}{6(a+b+c)}$$
$$= \frac{(b+c)(a^2 - b^2 + bc - c^2)}{12(a+b+c)} = 0,$$

so $a^2 - b^2 + bc - c^2 = 0 \Leftrightarrow \angle A = 60°$.

Note. The meaning of $B \cdot C$ depends on the choice of the origin. If $A = 0$, it represents $\overrightarrow{AB} \cdot \overrightarrow{AC} = \frac{b^2+c^2-a^2}{2}$; if $O = 0$, it represents $\overrightarrow{OB} \cdot \overrightarrow{OC} = \frac{\overrightarrow{OB}^2 + \overrightarrow{OC}^2 - a^2}{2} = \frac{2R^2 - a^2}{2}$; and so on.

Choosing the origin depends on the specific problem at hand, with the sole aim being simplification. Typically, the origin can be chosen as one of the vertices of a triangle, the circumcentre, incentre, or centroid of a triangle, or even the centre of a circle. It could also be a foot of an altitude or an intersection point of several lines within the figure. The choice should be made based on whichever option makes the solution more concise and avoids unnecessary complexity. For instance, in most cases, we prefer selecting vertex A of a triangle as the origin rather than the circumcentre O, as using the circumcentre introduces the radius R, which is often more complex than the side lengths a, b, and c. However, there are times when choosing the circumcentre becomes necessary, especially if the problem involves radii or for ease of representing the orthocentre. If a problem

Example 6. In the circumscribed quadrilateral $ABCD$, as shown in Figure 2.24, where $AB = a$, $BC = b$, $CD = c$, $DA = d$, and P is a free point in space, prove that

$$(ab + cd)(bcPA^2 + adPC^2)$$
$$= (bc + ad)(abPD^2 + cdPB^2).$$

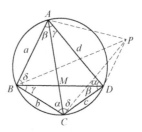

Figure 2.24

Proof. Let $P = 0$, which means we need to prove

$$(ab + cd)(bcA^2 + adC^2) = (bc + ad)(abD^2 + cdB^2).$$

Let M be the intersection point of AB and CD. Since $\alpha + \beta + \gamma + \delta = \pi$, we have

$$\frac{bcA + adC}{bc + ad} = M = \frac{dcB + abD}{dc + ab}.$$

Here, we used the co-side theorem. Therefore,

$$M^2 = \frac{bcA^2 + adC^2}{bc + ad} - \frac{abcd}{(bc + ad)^2}(A - C)^2$$
$$= \frac{dcB^2 + abD^2}{dc + ab} - \frac{abcd}{(dc + ab)^2}(B - D)^2.$$

We only need to prove

$$\frac{(A - C)^2}{(bc + ad)^2} = \frac{(B - D)^2}{(dc + ab)^2},$$

which is equivalent to

$$\frac{|AC|}{|BD|} = \frac{bc + ad}{dc + ab}.$$

Both sides of this equation are equal to $\frac{\sin(\alpha+\beta)}{\sin(\beta+\gamma)}$.

Corollary. When $P = A$, $AC = \sqrt{\frac{(bc+ad)(bd+ca)}{ab+cd}}$.

When $P = B$, $BD = \sqrt{\frac{(ab+cd)(bd+ca)}{bc+ad}}$.

Multiplying these two equations gives $AC \cdot BD = ac + bd$, which is Ptolemy's theorem.

Example 7. Let I be the incentre of $\triangle ABC$ and P be a free point. Prove that
$$PI^2 = \frac{aPA^2 + bPB^2 + cPC^2 - abc}{a+b+c}.$$

Proof. Let $P = 0$. We have
$$\left(\frac{aA + bB + cC}{a+b+c}\right)^2 - \frac{aA^2 + bB^2 + cC^2 - abc}{a+b+c}$$
$$= \frac{-abA^2 + 2abAB - abB^2 - acA^2 + a^2bc + ab^2c - bcB^2 + abc^2 + 2acAC + 2bcBC - acC^2 - bcC^2}{(a+b+c)^2}$$
$$= \frac{a^2bc - bc(B-C)^2 + a[-b(A-B)^2 - c(A-C)^2 + b^2c + bc^2]}{(a+b+c)^2}$$
$$= \frac{a^2bc - bca^2 + a(-bc^2 - cb^2 + b^2c + bc^2)}{(a+b+c)^2} = 0.$$

Alternatively, let $A = 0$. We have
$$(a+b+c)\left(P - \frac{aA+bB+cC}{a+b+c}\right)^2$$
$$- [a(P-A)^2 + b(P-B)^2 + c(P-C)^2 - abc]$$
$$= \frac{-b(a+c)B^2 + 2bcBC - (a+b)cC^2 + abc(a+b+c)}{a+b+c}$$
$$= \frac{-b(a+c)c^2 + bc(b^2 + c^2 - a^2) - (a+b)cb^2 + abc(a+b+c)}{a+b+c} = 0.$$

Note. Because P is irrelevant to the validity of the conclusion, it can be eliminated in the process. Therefore, by letting $A = 0$, the calculation can be further simplified. In the above result, when $P = I$, it corresponds to $aIA^2 + bIB^2 + cIC^2 = abc$.

Example 8. In Figure 2.25, O is the circumcentre of $\triangle ABC$. I is the incentre, and P and Q are the projections of I onto AB and BC, respectively. K and L are points on BA and BC, respectively, such that $BK = CQ$ and $BL = AP$. If BB' is the diameter of the circumcircle, prove that $B'I \perp KL$.

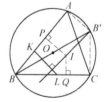

Figure 2.25

Proof 1. Let $O = 0$. Then,
$$K = \frac{\left(c - \frac{a+b-c}{2}\right)B + \frac{a+b-c}{2}A}{c},$$
$$L = \frac{\left(a - \frac{-a+b+c}{2}\right)B + \frac{-a+b+c}{2}C}{a},$$

Now, we can calculate the expression:
$$\left(-B - \frac{aA + bB + cC}{a+b+c}\right)(K - L)$$

$$= \frac{-2ABab(a+b) - a^2A^2(a+b-c) + B^2(a-c)(a+b+c)}{2ac(a+b+c)}$$

$$= \frac{-(2R^2 - c^2)ab(a+b) - a^2R^2(a+b-c) + R^2(a-c)(a+b+c)(a+2b+c)}{2ac(a+b+c)}$$

$$= 0.$$

Proof 2. $\overrightarrow{IB'} \cdot \overrightarrow{KL} = \overrightarrow{IB'} \cdot (\overrightarrow{KB} + \overrightarrow{BL}) = \overrightarrow{PA} \cdot \overrightarrow{KB} + \overrightarrow{QC} \cdot \overrightarrow{BL} = 0$.

Note. Proof 1 involves expressing B' in terms of B by setting $O = 0$, which leads to cumbersome calculations. It is not always necessary to explicitly express B'; instead, one can leverage perpendicular relationships for transformation. For a clever application of the vector method, readers can refer to [1]. From Proof 2, it can be observed that I is not necessarily the incentre. This problem illustrates that explicitly expressing all points and then proceeding with brute-force calculations can sometimes be tedious. It is possible to adopt clever methods, bypass point calculations, and achieve the objective without confrontation. The subsequent chapters introduce the method of identity, which follows a similar approach.

Example 9. Let point H be the orthocentre of an acute $\triangle ABC$ and P be any point in the plane. Prove that
$$(AP^2 - AH^2)\tan A + (BP^2 - BH^2)\tan B + (CP^2 - CH^2)\tan C$$
$$= PH^2 \tan A \tan B \tan C.$$

Proof. Let $P = 0$. Then,
$$H = \frac{\tan A \cdot A + \tan B \cdot B + \tan C \cdot C}{\tan A + \tan B + \tan C},$$
$(AP^2 - AH^2)\tan A = (A^2 - (A-H)^2)\tan A = (-H^2 + 2AH)\tan A.$

The left-hand side of the original expression becomes

$$-H^2(\tan A + \tan B + \tan C) + 2H(\tan A \cdot A + \tan B \cdot B + \tan C \cdot C)$$
$$= -H^2(\tan A + \tan B + \tan C) + 2H^2(\tan A + \tan B + \tan C)$$
$$= H^2 \tan A \tan B \tan C.$$

Note. The emphasis on an acute triangle in the problem allows for the use of the following identity:

$$\tan A + \tan B + \tan C = \tan A \tan B \tan C.$$

Example 10. In Figure 2.26, in $\triangle ABC$, with $AC > AB$, point D is taken on CA, M is the midpoint of AD, and N is the midpoint of BC. The extension of NM intersects the extension of BA at point E. Prove that $AE = AM \iff CD = AB$.

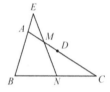

Figure 2.26

Proof 1. Let $A = 0$, $D = dC$, $M = \frac{D}{2}$, $N = \frac{B+C}{2}$,

$$E = t\frac{D}{2} + (1-t)\frac{B+C}{2} = \frac{1}{2}(1-t)B + \left(\frac{1-t}{2} + \frac{dt}{2}\right)C.$$

Solving $\frac{1-t}{2} + \frac{dt}{2} = 0$ gives $t = \frac{1}{1-d}$, $E = \frac{dB}{2(-1+d)}$,

$E^2 - M^2 = \frac{(B^2-(1-d)^2C^2)d^2}{4(1-d)^2}$,

$B^2 - (C-D)^2 = B^2 - (1-d)^2C^2$,

The proposition is proven.

Note. Here, $d \neq 1$ implies that D and C do not coincide; otherwise, $AB \parallel MN$ and E is at infinity.

Proof 2. Let $\overrightarrow{AD} = d\overrightarrow{AC}$, $\overrightarrow{AM} = \frac{\overrightarrow{AD}}{2}$, $\overrightarrow{AN} = \frac{\overrightarrow{AB}+\overrightarrow{AC}}{2}$,

$\overrightarrow{AE} = t\frac{\overrightarrow{AD}}{2} + (1-t)\frac{\overrightarrow{AB}+\overrightarrow{AC}}{2} = \frac{1}{2}(1-t)\overrightarrow{AB} + \left(\frac{1-t}{2} + \frac{dt}{2}\right)\overrightarrow{AC}$.

Solving $\frac{1-t}{2} + \frac{dt}{2} = 0$ gives $t = \frac{1}{1-d}$, $\overrightarrow{AE} = \frac{d\overrightarrow{AB}}{2(-1+d)}$,

$\overrightarrow{AE}^2 - \overrightarrow{AM}^2 = \frac{(\overrightarrow{AB}^2-(1-d)^2\overrightarrow{AC}^2)d^2}{4(1-d)^2}$,

$\overrightarrow{AB}^2 - (\overrightarrow{AC} - \overrightarrow{AD})^2 = \overrightarrow{AB}^2 - (1-d)^2\overrightarrow{AC}^2$.

The proposition is proven.

These two proof methods differ only in their presentation: one uses point geometry, whereas the other uses vector geometry. Upon comparison, it becomes evident that the point geometry representation is more concise.

In vector representation, you need to specify an origin point (A) and use arrow symbols, which can be cumbersome. Imagine a prolific writer who types slowly; it significantly impacts their writing because they are carrying a heavy burden.

It is crucial to emphasize that, due to the point geometry representation, it has not gained widespread popularity. It is still uncertain whether examiners would recognize it during assessments. Therefore, when answering questions, there are two options:

Option 1. Before presenting the solution, declare point A as the origin (the choice of origin depends on the specific problem; if not specified, assume any point, such as O, as the origin). Represent $\overrightarrow{AX}$ as X, $\overrightarrow{XY} = \overrightarrow{AY} - \overrightarrow{AX}$ as $Y - X$, and $\overrightarrow{AX} \cdot \overrightarrow{AY}$ as $X \cdot Y$ (it is essential not to omit the dot for the inner product, as it can lead to confusion). This approach is theoretically clear and can be adopted for regular problem-solving and academic writing. However, it may not be the safest choice for exams.

Option 2. Include the origin point and arrow symbols, completely rewriting everything in vector form. This approach is foolproof and highly recommended. Adding a few extra symbols will not significantly affect the time it takes, and it is well worth it to avoid losing marks.

Exercise 2.3

1. In Figure 2.27, in square $ABCD$, points E and F are located on segment BC and ray DC, respectively, with $2BE = EC$ and $2CF = DC$. The intersection of AE and BF is denoted as I. Prove that $IA \perp IC$.

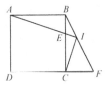

Figure 2.27

2. In $\triangle ABC$, D is a point on segment BC (distinct from B and C), and E and F are located on AB and AC, respectively, with $BD = CF$ and $CD = BE$. Point M is the midpoint of segment EF. Prove that $DM \perp EF \iff AB = AC$.

3. As shown in Figure 2.28, point O is the circumcentre of $\triangle ABC$, U is the bisector of angle $\angle A$ intersecting segment BC, and P is a point on segment AC. Prove that $UP \perp AO \iff AB = AP$.

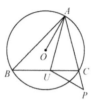

Figure 2.28

4. In Figure 2.29, in quadrilateral $ABCD$, $AB = AC$, $AD = DB$, $\angle BAD = 45°$, and $\angle BAC = 30°$. Prove that $DC \parallel AB$.

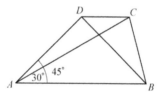

Figure 2.29

5. In Figure 2.30, in $\triangle ABC$, D is a point on an angle bisector of $\angle A$. Points E and F are located on sides AB and AC, respectively, such that $EC \parallel BD$ and $BF \parallel DC$. Points H and G are the midpoints of segments BF and CE, respectively. Prove that $HG \perp AD$.

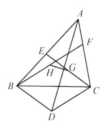

Figure 2.30

6. In Figure 2.31, in $\triangle ABC$, let AT be an angle bisector, M be the midpoint of BC, and ME be parallel to AT and intersects AB and AC (or their extensions) to points D and E, respectively. Prove that $BD = CE$.

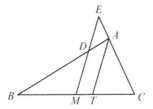

Figure 2.31

7. In Figure 2.32, in $\triangle ABC$, D is a point on BC. E and F are points on BC and AC, respectively. $AD \parallel EF$, line EF and BA intersects at G. Prove that D is the midpoint of $BC \iff EF + EG = 2AD$.

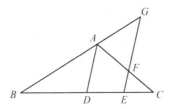

Figure 2.32

8. In $\triangle ABC$, the incentre is I, M is the midpoint of side AB, and CA is perpendicular to CB. Prove that $IM \perp IB$ if and only if the ratios of the sides are $BC : CA : AB = 3 : 4 : 5$.
9. In Figure 2.33, in $\triangle ABC$, AD, BE, and CF are angle bisectors. Prove that $DE \perp DF \iff \angle A = 120°$. (The 16th edition of the American Mathematics Olympiad, Enhanced Version, 1987)
10. In Figure 2.34, let P be a point on side BC of equilateral $\triangle ABC$. Draw AP and the perpendicular bisector of AP to intersect sides AB and AC at M and N, respectively. Prove that $BP \cdot PC = BM \cdot CN$. (The 1994 Anhui Province Mathematics Competition in China)

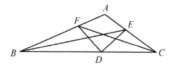

Figure 2.33

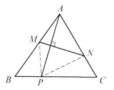

Figure 2.34

11. In Figure 2.35, in $\triangle ABC$, D is the midpoint of AC. E is the point where the incircle is tangent to AC, and F is the centre of the excircle opposite to vertex B. Prove that $BE \parallel DF$.

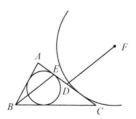

Figure 2.35

12. In Figure 2.36, in $\triangle ABC$, BE and CF are angle bisectors, O is the circumcentre, and I_A is the excentre opposite to $\angle A$. Prove that $OI_A \perp EF$.

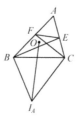

Figure 2.36

13. In Figure 2.37, AD is a median of $\triangle ABC$. P is a point on BC other than D. $PE \parallel AB$, $PF \parallel AC$, and BE intersects CF at O. OP intersects AD at K. Prove that quadrilateral $ABKC$ is a parallelogram.

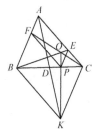

Figure 2.37

14. In Figure 2.38, AD is an angle bisector of $\triangle ABC$, where $AB \neq AC$. Lines DE and DF are parallel to BA and CA, respectively. Point G is the intersection of CF and BE. Prove that AG is perpendicular to BC if and only if AB is perpendicular to AC.

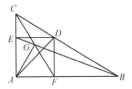

Figure 2.38

15. In Figure 2.39, P is a point on a median AD of $\triangle ABC$. BP intersects AC at E, and CP intersects AB at F. Prove that $BE = CF \iff AB = AC$.

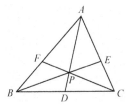

Figure 2.39

16. In Figure 2.40, in $\triangle ABC$, D and E are points on AB and AC, respectively. M is the midpoint of BE, N is the midpoint of CD, and P and Q are the intersections of MN with AB and AC, respectively. Prove that $BD = CE \iff AP = AQ$.

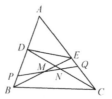

Figure 2.40

17. In Figure 2.41, in $\triangle ABC$, E and F are points on lines AB and AC, respectively. The midpoint of EF is P, and AP intersects BC at point D. Prove that if $AB = AC$, then $DE^2 - BE^2 = DF^2 - CF^2$.

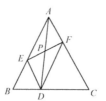

Figure 2.41

18. In Figure 2.42, in $\triangle ABC$, BL is an angle bisector, and BM is a median with $AB > BC$. $MD \parallel AB$ and $LE \parallel BC$. Prove that $ED \perp BL$.

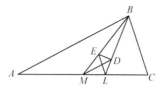

Figure 2.42

19. In $\triangle ABC$, AD is an angle bisector. Prove that $\frac{AB^2}{AD^2} = \frac{1}{2}\frac{BC}{DC} \iff BA \perp BC$.

20. In △ABC, the incircle touches AC at point M, and the incentre is I. Prove that $\frac{CI^2}{AI^2} = \frac{BC}{AB} \cdot \frac{MC}{AM}$.
21. In Figure 2.43, in △ABC, where $AB > AC$, $BD = CE$, $BL = AC$, CD intersects BE at F, and AT bisects ∠BAC. Prove that AT is parallel to LF.

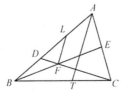

Figure 2.43

22. In acute △ABC, ∠A = 60°, and H, I, and O are the orthocentre, incentre, and circumcentre, respectively. Prove that $OI = IH$.
23. In △ABC, point I is the incentre, BD and CE are angle bisectors. Prove that $ID = IE \iff \angle B = \angle C$ or $\angle A = 60°$.
24. In Figure 2.44, PA is a tangent line and PB is drawn through the point P outside the circle O. Point C lies on PA, and point M is on BC. D is the intersection of PM and AB. Prove that M is the midpoint of CB if and only if $OD \perp BC$. (The 2006 Swiss National Team Selection Exam, Enhanced Edition)

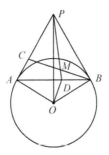

Figure 2.44

25. In Figure 2.45, in quadrilateral ABCD, AD intersects BC at E, and AC intersects BD at I. Prove that when AB is parallel to CD and $IC^2 = AI \cdot AC$, the centroid of △EDC coincides with the centroid of △IAB. (The 2005 Romanian Mathematical Olympiad)

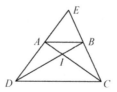

Figure 2.45

26. In Figure 2.46, in quadrilateral $ABCD$, two diagonals intersect at L, and M and N are the midpoints of AB and CD, respectively. Lines $AK \parallel CB$ and $KB \parallel AD$. Prove that $LK \parallel NM$.

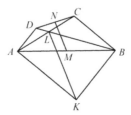

Figure 2.46

27. In Figure 2.47, K is a point inside $\triangle ABC$. Lines AK, BK, and CK intersect BC, CA, and AB at points D, E, and F, respectively. Lines AK, BK, and CK intersect EF, FD, and DE at points P, Q, and R, respectively. Lines AK, BK, and CK intersect QR, RP, and PQ at points X, Y, and Z, respectively. Prove that

$$\frac{1}{\frac{AD}{DX}+\frac{1}{2}} + \frac{1}{\frac{BE}{EY}+\frac{1}{2}} + \frac{1}{\frac{CF}{FZ}+\frac{1}{2}} = \frac{2}{3}.$$

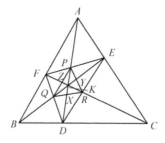

Figure 2.47

28. In Figure 2.48, in $\triangle ABC$, lines AD, BE, and CF intersect at point O. Line AD intersects line EF at P, line BE intersects line FD at Q, and line CF intersects line DE at R. Prove that

$$\frac{OP}{AP}\left(\frac{AF}{BF}+\frac{AE}{CE}\right)+\frac{OQ}{BQ}\left(\frac{BF}{AF}+\frac{BD}{CD}\right)+\frac{OR}{CR}\left(\frac{CD}{BD}+\frac{CE}{AE}\right)=2.$$

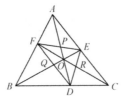

Figure 2.48

29. In Figure 2.49, in quadrilateral $ABCD$, AB intersects CD at E and AD intersects BC at F. J and K are the midpoints of AC and BD, respectively. G and H are the midpoints of DE and BF, respectively. Prove that $AI \parallel JK$.

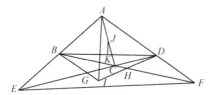

Figure 2.49

30. In Figure 2.50, points D and E are on the sides BC and AB of $\triangle ABC$, respectively. Line AD intersects CE at F, and line BF intersects DE at G. Parallel lines to BC through G intersect AB, CE, and AC at M, H, and N, respectively. Prove that $GH = NH$.

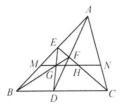

Figure 2.50

31. In Figure 2.51, I is the incentre of $\triangle ABC$. The incircle is tangent to BC at D. Let DE be a diameter of the incircle, and AE intersect BC at F. Prove that $BF = DC$.

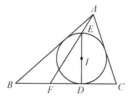

Figure 2.51

32. In Figure 2.52, $\triangle ABC$ has an inscribed circle $\odot I$. M is the midpoint of BC, and AH is an altitude. Line MI intersects AH at E. Prove that AE is equal to the radius of the incircle.

Figure 2.52

33. In Figure 2.53, $\triangle ABC$ has an incircle, which touches AB and BC at N and M, respectively. I is the incentre, and line CI intersects MN at P. Prove that $PA \perp PC$.

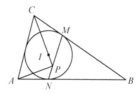

Figure 2.53

34. In Figure 2.54, P is a point inside $\triangle ABC$. Lines AP, BP, and CP intersect the opposite sides at D, E, and F, respectively. Points X, Y, and Z are the midpoints of EF, FD, and DE, respectively. Prove that the lines AX, BY, and CZ are concurrent.

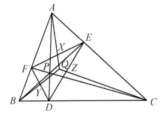

Figure 2.54

35. In Figure 2.55, in $\triangle ABC$, with $\angle B = 90°$, $AD = BC$, $CE = BD$, and CD intersects AE at G. Quadrilateral $DBFE$ is a rectangle. Prove that $FG \perp CD$.

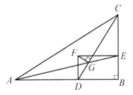

Figure 2.55

36. In Figure 2.56, in Rt$\triangle ABC$, where MN is a line parallel to BC, I is the incentre of $\triangle AMN$, and D is the midpoint of CM. Determine the angles of $\triangle BDI$.

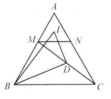

Figure 2.56

37. In Figure 2.57, point O is the circumcentre of $\triangle ABC$, where $AB \neq AC$. AU is an angle bisector, and I is a point on the extension of OA. Prove that $IU \perp BC \iff IA = IU$.

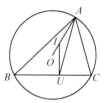

Figure 2.57

38. In Figure 2.58, E and F are points on AB and AC, respectively, with $BE = BC = CF$. BF intersects CE at K, and O is the circumcentre of $\triangle AEF$. Prove that $OK \perp BC$.

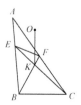

Figure 2.58

39. Given that the circumradius of $\triangle ABC$ is R, its circumcentre is O, the incentre is I, and the centroid is G and that $O \neq I$. Prove that $IG \perp BC$ if and only if $b = c$ or $b + c = 3a$. (The 2006 Romanian Mathematical Olympiad)

40. In Figure 2.59, O and I are the circumcentre and incentre of $\triangle ABC$, respectively, and D is on AC such that $DI \parallel AB$. Prove that $AB = AC \iff OD \perp CI$.

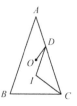

Figure 2.59

41. In Figure 2.60, $\odot O$, $\odot P$, and $\odot Q$ are the A-excircle, B-excircle, and C-excircle of $\triangle ABC$, respectively. Points D, E, F, G, H, and I are the points of tangency. The midpoints of segments EH, ID, and GF are X, Y, and Z, respectively. Prove that lines AZ and XY bisect each other. (Provided by Pan Chenghua)

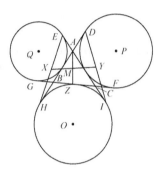

Figure 2.60

42. In Figure 2.61, in equilateral $\triangle ABC$, points D and E are on AB and BC, respectively, and $AD = BE$. AE intersects CD at F. Point H is on line BF, and EH intersects CD at G. If $EH = HG$. Prove that $AD = AG$.

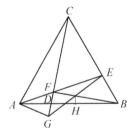

Figure 2.61

43. In Figure 2.62, I is the incentre of $\triangle ABC$. BI intersects AC at D. Construct $DE \parallel AI$ with E on BC, and EI intersects AB at F. Prove that $DF \perp AI \iff CA \perp CB$. (The Enhanced Edition of the 8th Northern Mathematical Olympiad Invitational Competition in 2012, China)

Figure 2.62

44. In Figure 2.63, in $\triangle ABC$, A_1, B_1, and C_1 are points on BC, CA, and AB, respectively, and AA_1, BB_1, and CC_1 meet at point P. Prove that P is the centroid of $\triangle ABC$ if and only if P is the centroid of $\triangle A_1B_1C_1$.

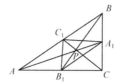

Figure 2.63

45. In Figure 2.64, let p be the semiperimeter of $\triangle ABC$. Points P and Q are chosen on rays BA and CA, respectively, such that $BP = CQ = p$. Let point K be the symmetric point A over the circumcentre $\triangle ABC$ and I be the incentre of $\triangle ABC$. Prove that $KI \perp PQ$. (2003 Silk Road International Math Olympiad)

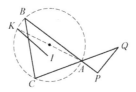

Figure 2.64

46. In Figure 2.65, given $\triangle ABC$, points M and N are the projections of B and C to bisectors of $\angle C$ and $\angle B$, respectively. (1) Prove that line MN intersects AC and AB at their points of contact with the incircle of $\triangle ABC$. (2) Prove that $2MN = BC \iff \angle A = 60°$. (The 2009 Sharygin Geometry Olympiad)

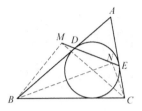

Figure 2.65

Reference

[1] Zhang Jingzhong and Peng Xicheng. *Clever Applications of the Vector Method*. Hubei Science and Technology Press, 2016.

Chapter 3
Identity-Based Method 1: Analytical Approaches

3.1 Classic Vector Identities

One-line proofs are often associated with inequalities, with the classic method being the sum of squares (SOS). During a past IMO competition, a participant famously earned a special award by providing a one-line proof of an inequality, showcasing both ingenious reasoning and brevity that impressed the judges. Can similar simplicity be achieved in plane geometry problems? Research in this area seems to be relatively scarce. However, after several years of investigation, we have discovered that it is indeed possible.

Due to habitual thinking patterns, geometry problems often give the impression of requiring deductive reasoning based on axioms, definitions, and theorems, with each step building upon the last. Conversely, algebraic problems involve various forms of splitting and equating quantities. In comparison, algebraic problems may seem to have more orderly and manageable solution steps, while geometry problems can sometimes be challenging to approach. Geometry and algebra do indeed have their own distinct characteristics, but they are not entirely mutually exclusive. Apart from the commonly used analytical methods, vector methods possess a unique blend of geometric and algebraic qualities. When applied effectively, they can make solving geometry problems as manageable as algebraic ones.

Example 1. In Figure 3.1, in rectangle $ABCD$, point E is the midpoint of AD, and a circle passing through points A, E, and C intersects the line CD at the other point F. Prove that AF is perpendicular to BE.

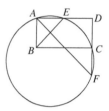

Figure 3.1

Proof.

$$\overrightarrow{EB} \cdot \overrightarrow{FA} = (\overrightarrow{DB} - \overrightarrow{DE}) \cdot (\overrightarrow{DA} - \overrightarrow{DF})$$

$$= (\overrightarrow{DA} + \overrightarrow{DC} - \frac{1}{2}\overrightarrow{DA}) \cdot (\overrightarrow{DA} - \overrightarrow{DF})$$

$$= \left(\frac{1}{2}\overrightarrow{DA} + \overrightarrow{DC}\right) \cdot (\overrightarrow{DA} - \overrightarrow{DF})$$

$$= \frac{1}{2}\overrightarrow{DA} \cdot \overrightarrow{DA} - \overrightarrow{DC} \cdot \overrightarrow{DF} + \overrightarrow{DA} \cdot \left(\overrightarrow{DC} - \frac{1}{2}\overrightarrow{DF}\right)$$

$$= 0.$$

Note. We use the fact that $\frac{1}{2}\overrightarrow{DA} \cdot \overrightarrow{DA} - \overrightarrow{DC} \cdot \overrightarrow{DF} = 0$ due to the circular power theorem, which states that A, E, C, and F are concyclic. $\overrightarrow{DA} \cdot \left(\overrightarrow{DC} - \frac{1}{2}\overrightarrow{DF}\right) = 0$ because the line DA is perpendicular to the line DC (DF).

The idea here is to first select points D, A, and C as reference points and represent the other points as much as possible using these reference points. For example, $\overrightarrow{DB} = \overrightarrow{DA} + \overrightarrow{DC}$, $\overrightarrow{DE} = \frac{1}{2}\overrightarrow{DA}$, and $\overrightarrow{DF} = t\overrightarrow{DC}$, with the aim of keeping it simple and avoiding the introduction of the parameter t. After converting the proof of the conclusion $\overrightarrow{EB} \cdot \overrightarrow{FA}$ into a representation based on reference points, to prove that the result is 0, we need to use the known conditions (the concyclicity and perpendicularity of the four points) to eliminate expressions.

In the above solution, point D repeatedly appears as the starting point for vectors, and writing vector symbols can be cumbersome. Therefore, it is necessary to introduce shorthand notation.

Conventionally, if O is taken as the origin, it is denoted as $O = 0$. $\overrightarrow{AB} = \overrightarrow{OB} - \overrightarrow{OA}$ is abbreviated as $B - A$, and $\overrightarrow{OA} \cdot \overrightarrow{OB}$ is abbreviated as AB (this is to be understood in context and not to be confused with line segment AB). If there is no assumed origin, any point can be set as the origin. This problem can be simplified into the form of an identity as follows.

Proof. Let

$$D = 0, \left(A + C - \frac{A}{2}\right)(A - F) - \left(A\frac{A}{2} - CF\right) - A\left(C - \frac{F}{2}\right) = 0.$$

It is worth emphasizing that since the current shorthand notation for point vectors has not yet been widely accepted, it might not be recognized in exams. Therefore, after generating the identity using point vectors, it is necessary to reintroduce the origin and vector symbols to return it to vector form. So, why not just write it directly in vector form? Imagine when you are thinking quickly: do you use a pen and paper, or do you use a chisel and a wooden board to write? Obviously, the former. Using convenient tools allows you to record as quickly as possible to keep up with the thought process. As for the need to carve on wooden boards, that can be considered at a later stage.

The reason for not writing it as $\left(A + C - \frac{A}{2}\right)(A - F) = \left(A\frac{A}{2} - CF\right) + A\left(C - \frac{F}{2}\right) = 0$ is to avoid special-casing the term $\left(A + C - \frac{A}{2}\right)(A - F)$. In fact, in mathematics, inverse propositions are often studied, which involve exchanging one term in the conclusion with a term in the conditions to see if the new proposition still holds. According to the obtained identity, it is evident that the original proposition can be transformed into three different propositions.

For example, in Figure 3.1, in parallelogram $ABCD$, where point E is the midpoint of AD and point F is another point on line DC, to prove the three conditions "$AF \perp BE$", "$DA \perp DC$", and "points A, E, C, and F are concyclic", knowing any two of them will yield the remaining.

As you can see, for some geometric problems that may initially appear complex, when you express the conclusions using the above method and then simplify and transform them, they naturally become equivalent to the given conditions in the problem. It is akin to expressing a geometric concept as an algebraic expression, and after equivalent transformations, it represents another geometric concept. It's like seeing ridges from one perspective and peaks from another.

Example 2. In Figure 3.2, in $\triangle ABC$, with BC as the longest side, points P and Q are taken on BC such that $BA = BQ$ and $CA = CP$. Prove that $PQ^2 = 2BP \cdot QC \iff AB \perp AC$.

Proof 1. Let $a = BC$, $b = CA = CP$, $c = BA = BQ$, $BP = a-b$, $PQ = -a+b+c$, and $QC = a-c$.

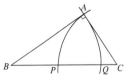

Figure 3.2

According to the identity $(-a+b+c)^2 - 2(a-b)(a-c) = b^2+c^2-a^2$, the proposition is proven.

Proof 2. Let $A = 0$. Then,
$$[(P-Q)^2 - 2(B-P)(Q-C)] - 2BC$$
$$+ [C^2 - (C-P)^2] + [B^2 - (B-Q)^2] = 0.$$

The representation as universal vectors is
$$[(\overrightarrow{AP} - \overrightarrow{AQ})^2 - 2(\overrightarrow{AB} - \overrightarrow{AP}) \cdot (\overrightarrow{AQ} - \overrightarrow{AC})] - 2\overrightarrow{AB} \cdot \overrightarrow{AC}$$
$$+ [\overrightarrow{AC}^2 - (\overrightarrow{AC} - \overrightarrow{AP})^2] + [\overrightarrow{AB}^2 - (\overrightarrow{AB} - \overrightarrow{AQ})^2] = 0.$$

In the traditional study of geometry, it is not easy to construct an identity with geometric meaning, as in the case of Proof 1 in Example 2. However, if you explore the treasure trove of vector identities, it becomes easy to construct identities with geometric meaning. This discovery builds a bridge between numbers and shapes, merging them further, as algebraic identities can have geometric meanings and geometric problems can be used to construct algebraic identities. In fact, vector identities have appeared in secondary school mathematics for some time; they just have not been given much attention.

A few classic examples are as follows:

1. The cosine rule in $\triangle ABC$: $(C-A)^2 - (A-B)^2 - (B-C)^2 + 2(B-A)(B-C) = 0$, which is a special case of the Pythagorean theorem: $BA \perp BC \iff b^2 = c^2 + a^2$.
2. If we treat the identity $(a+b)^2 - (a-b)^2 = 4ab$ with a and b as vectors, where $a = \overrightarrow{OA}$ and $b = \overrightarrow{OB}$, then $\left(\overrightarrow{OA} + \overrightarrow{OB}\right)^2 - \left(\overrightarrow{OA} - \overrightarrow{OB}\right)^2 = 4\overrightarrow{OA} \cdot \overrightarrow{OB}$ holds geometric significance: in a parallelogram, if one angle is a right angle, then the diagonals are of equal length, and vice versa.
3. The difference of squares formula: $A^2 - B^2 = (A+B)(A-B)$ also has a geometric interpretation: in a parallelogram, the adjacent sides are equal if and only if the diagonals are perpendicular.

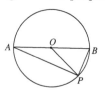

Figure 3.3

4. Identity: $(P-A)(P-B) = \left(P - \frac{A+B}{2}\right)^2 - \left(\frac{A-B}{2}\right)^2$. This represents the theorem of Thales: if point P satisfies $\left(P - \frac{A+B}{2}\right)^2 = \left(\frac{A-B}{2}\right)^2$, then $\angle APB$ is a right angle, and vice versa. As shown in Figure 3.3, O is the midpoint of AB, so $\angle APB = 90° \Leftrightarrow OA = OP$.

Example 3. In Figure 3.4, in $\triangle ABC$, D is a point on BC. If AB is perpendicular to AC and AD is perpendicular to BC, prove that $AB^2 = BC \cdot BD$ and $AD^2 = BD \cdot DC$ (the right-angled triangle projection theorem).

Proof.

$$(A-B)^2 - (B-C)(B-D) = (A-B)(A-C) - (B-C)(A-D).$$

This actually proves that in $\triangle ABC$, if D is a point on BC and any two of the following three conditions are known: AB is perpendicular to AC, AD is perpendicular to BC, and $AB^2 = BC \cdot BD$, then the remaining condition holds. In other words, one identity simultaneously proves three propositions. This fact will not be mentioned further in the following text:

$$(A-D)^2 - (B-D)(D-C) = (A-B)(A-C) + (D-A)(D-C)$$
$$+ (D-A)(D-B).$$

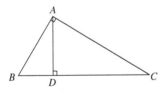
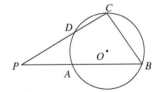

Figure 3.4　　　　　　　Figure 3.5

Example 4. In Figure 3.5, quadrilateral $ABCD$ is inscribed in circle O, and line AB intersects CD at point P. Prove that $PA \cdot PB = PC \cdot PD$ (the circular power theorem).

Proof. Let $O = 0$, then

$$(P-A)(P-B) - (P-C)(P-D) - 2\left(\frac{A+B}{2}\right)(B-P)$$
$$+ 2\left(\frac{C+D}{2}\right)(D-P) - (D^2 - B^2) = 0.$$

3.2 Examples: Identities with Two Terms

Example 1. In Figure 3.6, in $\triangle ABC$, $AB = AC$. Extend AB to E such that $BE = AB$. D is the midpoint of AB. Prove that $CE = 2CD$.

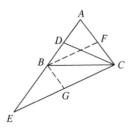

Figure 3.6

Proof 1. Let $A = 0$. Then,

$$(C - 2B)^2 - 4\left(C - \frac{B}{2}\right)^2 - 3(B^2 - C^2) = 0.$$

In vector notation,

$$(\overrightarrow{AC} - 2\overrightarrow{AB})^2 - 4\left(\overrightarrow{AC} - \frac{\overrightarrow{AB}}{2}\right)^2 - 3(\overrightarrow{AB}^2 - \overrightarrow{AC}^2) = 0.$$

Traditional Euclidean methods involve adding auxiliary lines and constructing congruent or similar triangles for solving such problems. There are many approaches to adding auxiliary lines in this case, such as constructing a median BF of $\triangle ABC$ or a median BG of $\triangle BEC$.

Proof 2. Construct a median BF of $\triangle ABC$. Then, $CE = 2BF$. It can be easily proved that $\triangle DBC \cong \triangle FCB$, which implies $BF = CD$. Thus, $CE = 2BF = 2CD$.

Note. In Proof 1, we use $2B$ to indicate that $BE = AB$ and $\frac{B}{2}$ to denote that D is the midpoint of AB. The condition $3(B^2 - C^2) = 0$ is used because $AB = AC$. Each condition is used only once, and no additional auxiliary lines are needed, making the solution straightforward. The only extra step is changing $CE = 2CD$ to $CE^2 = 4CD^2$. In Proof 2, the first challenge is adding auxiliary lines. Currently, there is no universally applicable method for adding auxiliary lines; the process often relies on experience and sometimes even luck. Whether or not auxiliary lines are added, the propositions still hold, demonstrating that auxiliary lines do not fundamentally affect the problem but are merely tools for solving it. To prove $\triangle DBC \cong \triangle FCB$, we need to show that $DB = FC$, which relies on the midpoint information $AD = DB$, $AF = FC$, and $AB = AC$. We also need to prove $\angle ABC = \angle ACB$, indicating that $AB = AC$. The condition $BC = CB$ is not specific to this problem and is a universally accepted equality. The addition of the median BF is an ingenious construction that is used twice. $AB = AC$ is used twice, and the identity $BC = CB$ (symmetric equality) is used once. Ideally, in terms of simplicity, if the conditions

Identity-Based Method 1: Analytical Approaches

in the problem are neither too few nor too many, the best scenario is to use each condition only once. It is advisable to minimize the use of information that is outside of the problem statement.

Example 2. In Figure 3.7, circle O is inscribed in quadrilateral $ABCD$. E, F, and G are the midpoints of BC, AD, and CD, respectively. Construct a parallelogram $FOEN$. Prove that GN is perpendicular to AB (Cantor's theorem).

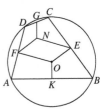

Figure 3.7

Proof. Let $O = 0$. Then,

$$2\left(\frac{A+D}{2} + \frac{B+C}{2} - \frac{C+D}{2}\right) \cdot (A-B) + (B^2 - A^2) = 0.$$

In vector notation,

$$2\left(\frac{\overrightarrow{OA} + \overrightarrow{OD}}{2} + \frac{\overrightarrow{OB} + \overrightarrow{OC}}{2} - \frac{\overrightarrow{OC} + \overrightarrow{OD}}{2}\right)$$
$$\cdot (\overrightarrow{OA} - \overrightarrow{OB}) + (\overrightarrow{OB}^2 - \overrightarrow{OA}^2) = 0.$$

Note. The solver usually assumes that the given conditions are not redundant and often uses all the conditions provided by the proposition. However, when using the method of identities, representing multiple conditions as polynomial equations to derive the conclusion, it becomes easier to discover that some conditions are redundant. From the identity, it is evident that the condition $OA = OB = OC = OD$ is too strong, and it is sufficient to ensure that $OA = OB$. In fact, A, B, C, D, and O do not need to lie on the same plane or even within the same three-dimensional space. These five points can define a four-dimensional space. This can be modified as follows: in quadrilateral $ABCD$, $OA = OB$, and K, E, F, and G are the midpoints of AB, BC, AD, and CD, respectively. Construct a parallelogram $FOEN$. Prove that GN is perpendicular to AB and that quadrilateral $GNKO$ is a parallelogram.

Example 3. In Figure 3.8, within parallelogram $ABCD$, extend AB on both sides to points E and F such that $AE = AB = BF$. Connect CE and DF. Prove that $AD = 2AB \iff EC$ is perpendicular to FD.

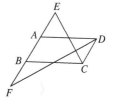

Figure 3.8

Proof. Let $A = 0$. Then,
$$(-B - (B+D))(2B - D) + (4B^2 - D^2) = 0.$$
In vector notation,
$$(-\overrightarrow{AB} - (\overrightarrow{AB} + \overrightarrow{AD})) \cdot (2\overrightarrow{AB} - \overrightarrow{AD}) = -(4\overrightarrow{AB}^2 - \overrightarrow{AD}^2).$$

Example 4. In Figure 3.9, within $\triangle ABC$, CA is perpendicular to CB. AD and BE are medians. Prove that
$$AD^2 + BE^2 = \frac{5}{4}AB^2.$$

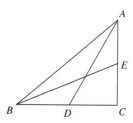

Figure 3.9

Proof.
$$\left[\left(A - \frac{B+C}{2}\right)^2 + \left(B - \frac{A+C}{2}\right)^2 - \frac{5}{4}(A-B)^2\right]$$
$$- \frac{1}{2}(C-A)(C-B) = 0.$$

Example 5. In Figure 3.10, in trapezium $ABCD$, AD is parallel to BC. Take a point E on the longer base BC such that BE equals the length of the trapezoid's middle line. Prove that AC is perpendicular to BD if and only if DE also equals the length of that middle line. (Enhanced Version of the 17th Russian Mathematical Olympiad Problems, 1997)

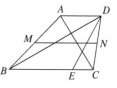

Figure 3.10

Proof.
$$\left[\left(\frac{C+D}{2} + B - \frac{A+B}{2} - D\right)^2 - \left(\frac{A+B}{2} - \frac{C+D}{2}\right)^2\right]$$
$$+ (A-C)(B-D) = 0.$$

Note. The quadrilateral $BENM$ is a parallelogram.

Example 6. Given that point O is any point on the plane of rectangle $ABCD$, prove that $OA^2 + OC^2 = OB^2 + OD^2$.

Proof. Let $O = 0$. Then, $D = A + C - B$, and
$$[A^2 + C^2 - B^2 - (A + C - B)^2] + 2(B - A)(B - C) = 0.$$

Note. This proposition has broad applications. O does not necessarily have to lie on the plane of rectangle $ABCD$. From the above identity, we can derive the following: in quadrilateral $ABCD$, with O being any point, "$OA^2 + OC^2 = OB^2 + OD^2$", "$BA$ is perpendicular to BC", and "$ABCD$ is a parallelogram"; if you know any two of these conditions, you can prove the remaining.

Example 7. In space quadrilateral $ABCD$, $AD^2 + BC^2 = AB^2 + CD^2 \iff AC \perp BD$.

Proof. $(A - D)^2 + (B - C)^2 - (A - B)^2 - (C - D)^2 = 2(A - C)(B - D).$

Example 8. In isosceles trapezium $ABCD$ with AB parallel to CD, prove that $AC^2 = AD^2 + AB \cdot DC$.

Proof. $[(A - C)^2 - (A - D)^2 - (A - B)(D - C)] + \left[\left(\frac{A+B}{2} - D\right)^2 - \left(\frac{A+B}{2} - C\right)^2\right] = 0.$

Example 9. In $\triangle ABC$ with $AB = AC$, P is a point on BC. Prove that $AB^2 = AP^2 + BP \cdot PC$.

Proof. Let $A = 0$. Then, $B^2 - P^2 - (P - B)(C - P) - (B + C)(B - P) = 0.$

Note. The geometric significance of $(B + C)(B - P) = 0$ is that BP is perpendicular to a median on BC.

Example 10. Point S lies on the circumcircle O of $\triangle ABC$, and H is the orthocentre. Prove that the midpoint K of SH lies on the nine-point circle of $\triangle ABC$.

Proof. Let $O = 0$, then $H = A + B + C$, and the centre of the nine-point circle is $\frac{A+B+C}{2}$:
$$\left(\frac{A+B+C}{2} - \frac{B+C}{2}\right)^2 - \left(\frac{A+B+C}{2} - \frac{S+A+B+C}{2}\right)^2$$
$$-\frac{1}{4}(A^2 - S^2) = 0.$$

Example 11. In Figure 3.11, in trapezium $ABCD$ with AD parallel to BC and $\frac{BC}{AD} = \frac{1}{3}$, point M is on side CD, and $\frac{CM}{MD} = \frac{2}{3}$. Prove that $AB = AD \iff BD \perp AM$. (The 2018 Ukraine Geometry Olympiad, Adapted Version)

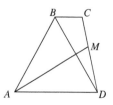

Figure 3.11

Proof.

$$\frac{5}{3}(B - D)\left(A - \frac{3\left(B - \frac{A-D}{3}\right) + 2D}{5}\right)$$
$$+ [(A - B)^2 - (A - D)^2] = 0.$$

Note. $\frac{A-D}{3} = B - C$, and $M = \frac{3C + 2D}{5} = \frac{3\left(B - \frac{A-D}{3}\right) + 2D}{5}$.

Example 12. In Figure 3.12, in circumscribed quadrilateral $ACBD$ with mutually perpendicular diagonals, E is the intersection point of diagonals AB and CD, and F is the midpoint of AC. Prove that $EF \perp BD$.

Proof. Let the centre of the circle be $O = 0$, $E = \frac{A+B}{2} + \frac{C+D}{2} = \frac{A+B+C+D}{2}$, and $F = \frac{A+C}{2}$:

$$2\left(\frac{A+B+C+D}{2} - \frac{A+C}{2}\right)(B - D) + (D^2 - B^2) = 0.$$

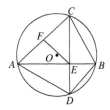

Figure 3.12

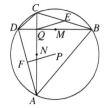

Figure 3.13

Example 13. In Figure 3.13, there is a cirumscribed quadrilateral $ABCD$ within a circle with centre P. The diagonals intersect at point Q, and E and F are the midpoints of BC and AD, respectively. Prove that $QE = PF$.

Proof. Let the centre of the circle be $P = 0$, and let M and N be the midpoints of BD and AC, respectively. Then,

$$Q = M + N = \frac{A+B+C+D}{2},$$

$$\left(\frac{A+B+C+D}{2} - \frac{B+C}{2}\right) - \frac{A+D}{2} = 0.$$

Note. This shows that QE and FP are parallel and equal in length.

Example 14. In Figure 3.14, within $\triangle ABC$, O is the circumcentre, H is the orthocentre, and $\angle C = 60°$. N is the midpoint of the minor arc AB of the circumcircle of $\triangle ABC$. Prove that $CN \perp OH$. (1994 Bulgarian Competition)

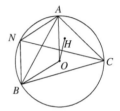

Figure 3.14

Proof. Construct an equilateral $\triangle ABD$, with C and D on the same side of AB. Let $O = 0$, $H = A + B + C$, $N = A + B$, and $D = -A - B$. Then, $(A + B - C)(A + B + C) + [C^2 - (A+B)^2] = 0$.

Example 15. In Figure 3.15, within $\triangle ABC$, where $AB = AC$, point D is the midpoint of BC, AD is connected, point E is the midpoint of AD, and DG is perpendicular to BE at point G. Point F is the midpoint of AC. Prove that $GF = DF$.

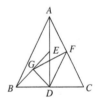

Figure 3.15

Proof.

$$\left[\left(G - \frac{A+C}{2}\right)^2 - \left(\frac{B+C}{2} - \frac{A+C}{2}\right)^2\right]$$

$$+ 2\left(G - \frac{B+C}{2}\right)\frac{\frac{B+C}{2} + A}{2} - \frac{B+G}{2} = 0.$$

Note. It is easy to see that $GF = DF$ does not depend on $AB = AC$. Traditional methods might inadvertently use this condition, but the method using identities helps identify redundant conditions.

Exercise 3.2

1. In Figure 3.16, in parallelogram $ABCD$, P is a point such that $PC = BC$. Prove that the line MN, connecting the midpoints of AP and CD is perpendicular to BP.
2. In Figure 3.17, in $\triangle ABC$, two medians AD and BE intersect at G, and BA is extended to F such that $AB = AF$. Prove that $FG = AC \iff AD \perp BE$.
3. In Figure 3.18, in $\triangle ABC$, where $AB = 3BC$, P and Q are two points on side AB such that $PQ = AP = QB$, and point M is the midpoint of AC. Prove that $\angle PMQ = 90°$. (2018–2019 Polish Junior Secondary School Math Olympiad Plane Geometry Problem)

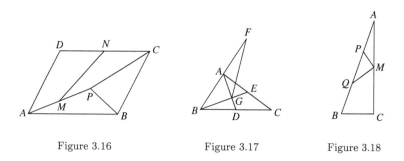

Figure 3.16   Figure 3.17   Figure 3.18

4. In Figure 3.19, on the extension of side AB of $\triangle ABC$, a point D is taken such that $BD = AB$. Connect DC, and choose a point E on side BC so that $CE = 2EB$. If $3AE = CD$, prove that $\triangle ABC$ is a right-angled triangle.

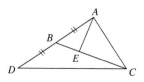

Figure 3.19

5. In Figure 3.20, in $\triangle ABC$, D is the midpoint of AC and E is a point on side BC such that $2BE = EC$. Prove that $AB \perp AC \iff AE = \frac{2}{3}BD$.

Identity-Based Method 1: Analytical Approaches   73

6. In Figure 3.21, in quadrilateral $ABCD$, $AD \parallel BC$ and $BD \perp AC$. Let M and N be the midpoints of AD and BC, respectively. Prove that $2MN = AD + BC$.

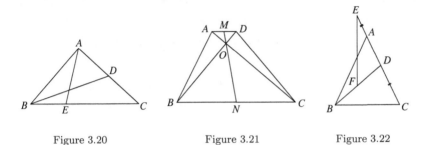

Figure 3.20   Figure 3.21   Figure 3.22

7. In Figure 3.22, in $\triangle ABC$, where $AB = AC$, points D and E are located on side CA and its extension, respectively, such that $CD = 2AE$. Point F is the midpoint of BD. Prove that $EF \perp BC$.
8. In tetrahedron $ABCD$, M is the midpoint of BC, N is the midpoint of AC, P is the midpoint of DA, and Q is the midpoint of DB. If $AB = CD$, prove that $PM \perp QN$.

3.3 Examples: Identities with Multiple Terms

Example 1. In Figure 3.23, it is known that PQ is parallel to the side AD of rectangle $ABCD$. Prove that $AP^2 + CQ^2 = PB^2 + DQ^2$.

Proof. $[(A - P)^2 + (C - Q)^2 - (P - B)^2 - (A + C - B - Q)^2] + 2(B - A)(B - C) - 2(B - A)(P - Q) = 0$.

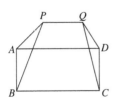
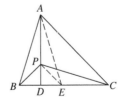

Figure 3.23   Figure 3.24

Example 2. In Figure 3.24, within $\triangle ABC$, $AB < AC$, and AD is an altitude on side BC. Point P lies on AD and satisfies $\angle ABP = \angle ACP$. Prove that P is the orthocentre of $\triangle ABC$.

Proof. Let $D = 0$. We have $(P-B)(A-C) - (PA+BC) + PC + AB = 0$.

Note. Let E and B be symmetric about line AD. Given that $AB < AC$ and $\angle ABP = \angle ACP$, we can deduce that the four points E, C, A, and P are concyclic.

Example 3. In Figure 3.25, in $\triangle ABC$, $AB = AC$, D is a point on BC, and E is a point on AB. If $AD \perp AC$ and $DB = DE$, prove that the points A, E, D, and C are concyclic.

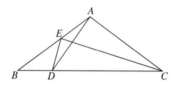

Figure 3.25

Proof. Let $B = 0$. We have $(AE - CD) - 2D\left(A - \frac{C}{2}\right) + 2A\left(D - \frac{E}{2}\right) = 0$.

Example 4. In Figure 3.26, H is the orthocentre of $\triangle ABC$, and AD is an altitude. Point E is symmetric to point H about side BC. Prove that point E lies on the circumcircle of $\triangle ABC$.

Proof. Let $D = 0$. We have $(-HA - BC) - (B-H)(A-C) + HC + AB = 0$.

Note. The expression of the conclusion is $-HA - BC$. To eliminate HA and BC, we introduce the expression $(B-H)(A-C) = 0$. This introduces HC and AB, but both of these terms equal 0.

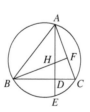 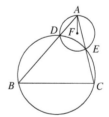

Figure 3.26   Figure 3.27

Identity-Based Method 1: Analytical Approaches

Example 5. In Figure 3.27, in $\triangle ABC$, points D and E lie on sides AB and AC, respectively, such that points B, C, E, and D are concyclic. F is the circumcentre of $\triangle ADE$. Prove that $AF \perp BC$.

Proof. Let $A = 0$. We have $F(B-C) - \left(F - \frac{D}{2}\right)B + \left(F - \frac{E}{2}\right)C - \frac{1}{2}(BD - CE) = 0$.

Note. To eliminate FB and FC from the conclusion $F(B - C)$, we introduce the conditions $\left(F - \frac{D}{2}\right)B = 0$ and $\left(F - \frac{E}{2}\right)C = 0$. The added BD and CE can obviously be eliminated using the circular power theorem, $BD - CE = 0$.

Example 6. In Figure 3.28, in $\triangle ABC$, a line passing through A intersects a circle with BC as a diameter at point D. Let E be a point on AB such that $AD = AE$. A perpendicular line is drawn from E to AB, intersecting AC at point F. Prove that $\frac{AE}{AB} = \frac{AC}{AF}$.

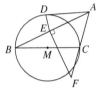

Figure 3.28

Proof. Let $A = 0$. We have $(EF - BC) + (D^2 - E^2) + 2D\left(\frac{B+C}{2} - D\right) + (D - B)(D - C) + E(E - F) = 0$.

Note. To eliminate EF and BC from the conclusion $EF - BC$, we consider introducing $(D - B)(D - C) = 0$ and $E(E - F) = 0$. (We choose not to introduce a new term such as $(E - B)(E - F) = 0$ because we would rather avoid the new term EB, if possible.) The addition of $D^2 - E^2 = 0$ can be used to eliminate D^2 and E^2, while eliminating BD and CD can be handled by introducing $D\left(\frac{B+C}{2} - D\right) = 0$.

Example 7. In Figure 3.29, K is the midpoint of the circumcentre O and the orthocentre H of $\triangle ABC$. Point L lies on the perpendicular bisector of side AC such that $BO \perp BL$, and point M lies on the perpendicular bisector of side AB such that $CO \perp CM$. Prove that $AK \perp LM$ (The 2016 Balkan Mathematical Olympiad).

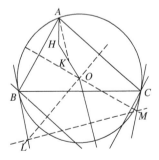

Figure 3.29

Analysis. When dealing with the circumcentre and orthocentre, it is often necessary to make use of the properties of the Euler line. Let $O = 0$, then $H = A + B + C$.

This helps in representing the geometric relationships. To express the conclusion $AK \perp LM$, we write $2\left(A - \frac{A+B+C}{2}\right)(L-M) = AL - BL - CL - AM + BM + CM$, which upon multiplying by 2 makes the expression an integral expression, and when expanded, it becomes easy to note that $AL - CL = L(A-C) = 0$, where the vector representation is $\overrightarrow{OL} \cdot \overrightarrow{AC} = 0$. Similarly, $-AM + BM = (B-A)M = 0$. Now, we only need to prove $CM - BL = 0$. Adding and subtracting terms, we get $CM - BL + B^2 - C^2 = C(M-C) + B(L-B) = 0$. Here, $C(M-C) = 0$ has the vector representation $\overrightarrow{OC} \cdot \overrightarrow{MC} = 0$, which is due to the properties of tangents. Similarly, $B(L-B) = 0$. $B^2 - C^2 = 0$ because the radii are equal. In summary, we eliminate each term of the conclusion step by step, sometimes utilizing the "add and subtract" technique.

Proof. Let $O = 0$, then $H = A + B + C$, $2\left(A - \frac{A+B+C}{2}\right)(L-M) - L(A-C) + B(L-B) + M(A-B) + C(C-M) + (B^2 - C^2) = 0$.

Example 8. In Figure 3.30, H is the orthocentre of $\triangle ABC$, and M and N are the midpoints of BC and AH, respectively. Circles with diameters AH and BC intersect at point I. Prove that $IM \perp IN$.

Proof. Let $H = 0$, then we have $4\left(I - \frac{B+C}{2}\right)\left(I - \frac{A}{2}\right) - 2(I-A)I - 2(I-B)(I-C) + C(B-A) + B(C-A) = 0$.

Note. The difficulty in this problem lies in how to represent point I. We can ignore the circles for now and focus on the conditions that point I must satisfy:
$$(I-A)(I-H) = 0, \quad (I-B)(I-C) = 0.$$

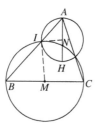

Figure 3.30

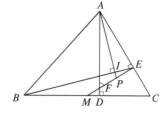

Figure 3.31

Example 9. In Figure 3.31, in acute $\triangle ABC$, where $AB > AC$, M is the midpoint of BC, and $AD \perp BC$ at D. The perpendicular from M to line AC intersects AC and AD at points E and F, respectively. P is the midpoint of EF. Prove that $AP \perp BE$.

Proof. Let $E = 0$, then we have $2B\left(A - \frac{F}{2}\right) - 2A\frac{B+C}{2} + CF + (F - A)(B - C) = 0$.

Note. To simplify the equation, we aim to represent points using as few variables as possible and avoid introducing unnecessary points. In this problem, $M = \frac{B+C}{2}$ and $P = \frac{E+F}{2}$. The purpose of point D is primarily to locate F. If the condition $(F - A)(B - C) = 0$ is given, then the purpose of D is fulfilled, and it can be removed without affecting the proof. Now, we are left with points A, B, C, E, and F. Since E is the intersection point of two perpendicular lines, setting $E = 0$ simplifies the equation. With this, we aim to prove that $B\left(A - \frac{F}{2}\right) = 0$, for which the crucial condition used is $(F - A)(B - C) = 0$. Combining this with $FC = 0$ and $A\frac{B+C}{2} = 0$, we can express $B\left(A - \frac{F}{2}\right) = 0$, and by completing the square, it becomes an identity.

Example 10. Points P and Q are on sides AB and AC of $\triangle ABC$, respectively, where $AB \perp AC$. M is the midpoint of side BC, and $PM \perp QM$. Prove that
$$PB^2 + QC^2 = PM^2 + QM^2.$$

Proof.
$$\left(\frac{B+C}{2} - P\right)^2 + \left(\frac{B+C}{2} - Q\right)^2 - (P - B)^2 - (C - Q)^2$$
$$+ 2\left(P - \frac{B+C}{2}\right)\left(Q - \frac{B+C}{2}\right) - 2(P - B)(Q - C) = 0.$$

Note. First, express $PB^2 + QC^2 = PM^2 + QM^2$ and $PM \perp QM$ clearly. It is observed that these expressions do not involve point A. So, we choose $(P - B)(Q - C)$ to represent $AB \perp AC$. This could lead to nine possible expressions when combining $AB \perp AC$, with points A, B, and P being collinear and points A, C, and Q being collinear (e.g., $(P - A)(Q - C) = 0$, $(B - A)(Q - C) = 0$, etc.). Selecting the required expressions can be cumbersome.

Example 11. In Figure 3.32, point D inside equilateral $\triangle ABC$ satisfies $\angle ADC = 150°$. Prove that a triangle with side lengths $|AD|$, $|BD|$, and $|CD|$ is a right-angled triangle (2003 Nordic Mathematical Competition).

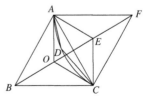

Figure 3.32

Proof. Let $B = O$, then $[D^2 - (D - A)^2 - (D - C)^2] + [(A - C)^2 - C^2] + [(A + C - D)^2 - A^2] = 0$.

Note. The key to solving this problem lies in the application of the condition $\angle ADC = 150°$. This requires constructing a parallelogram and using properties such as the inscribed angle theorem. Construct a parallelogram $ABCF$. Based on the condition, point D lies on the circle with centre F and radius FA. From the identity, we can see that point D does not necessarily have to be inside $\triangle ABC$ or even on the plane containing $\triangle ABC$. It only needs to satisfy $FD = BA$. The identity method can capture the essence of the problem and easily extend the proposition, especially when extending it to three-dimensional space. This is not elaborated on in other problems.

Example 12. In Figure 3.33, in rhombus $ABCD$, point E is chosen on BC such that $AB = AE$. Line AC intersects DE at F. Prove that points A, F, E, and B are concyclic.

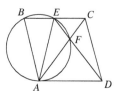

Figure 3.33

Proof. Let $C = 0$, then we have

$$(BE - AF) - 2F\left(B - \frac{A}{2}\right) + 2B\left(F - \frac{E}{2}\right) = 0.$$

Note. The reason for $B\left(F - \frac{E}{2}\right) = 0$ is that $AB = AE = DC$, so $FE = FC$. Writing the condition $AB = AE$ directly as $(A - B)^2 - (A - E)^2 = 0$ would introduce the term E^2, making it harder to eliminate. The goal is to use the condition effectively to eliminate the conclusion $BE - AF$.

Example 13. In Figure 3.34, given cyclic quadrilateral $ABCD$, DP is parallel to CA, intersecting the extension of BA at P. Prove that

$$DA \cdot DC = AP \cdot BC.$$

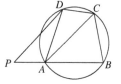

Figure 3.34

Proof. Let the centre of the circle be $O = 0$, then we have

$$[(D - A)(D - C) + (A - P)(B - C)]$$
$$- 2\frac{A + C}{2}(P - D) + 2\frac{A + B}{2}(P - A) + (A^2 - D^2) = 0.$$

Note. $\angle ADC$ and $\angle ABC$ are complementary. Expressing the given conditions and conclusions in terms of point geometry is the first step in establishing the identity. Without this foundational step, it is impossible to proceed. Therefore, utilizing the given information effectively is crucial.

Example 14. In Figure 3.35, given cyclic quadrilateral $ABCD$, the diagonal AC intersects BD at E, and $BE = DE$. Prove that $AB^2 + BC^2 + CD^2 + DA^2 = 2AC^2$. (the 2019 Beijing Mathematical Competition)

Proof.
$$[(A-B)^2 + (B-C)^2 + (C-D)^2 + (D-A)^2 - 2(A-C)^2]$$
$$-4\left[\left(\frac{B+D}{2} - A\right)\left(\frac{B+D}{2} - C\right)\right.$$
$$\left.-\left(\frac{B+D}{2} - B\right)\left(\frac{B+D}{2} - D\right)\right] = 0.$$

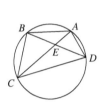

Figure 3.35

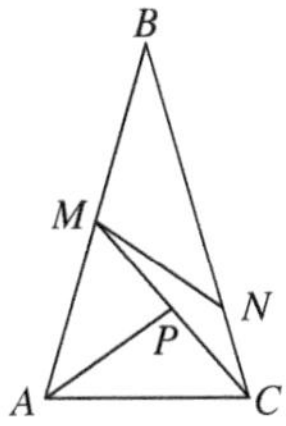

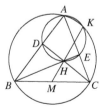

Figure 3.36

Example 15. In Figure 3.36, in $\triangle ABC$, where $AB = BC$, M is the midpoint of AB, and P is the midpoint of CM. Point N divides side BC into a $3:1$ ratio (starting from vertex B). Prove that $AP = MN$. (2013 Moscow Mathematical Olympiad)

Proof. Let $B = 0$, then we have
$$\left(A - \frac{\frac{A}{2}+C}{2}\right)^2 - \left(\frac{A}{2} - \frac{3C}{4}\right)^2 - \frac{5}{16}(A^2 - C^2) = 0.$$

Example 16. In Figure 3.37, two altitudes CD and BE of $\triangle ABC$ intersect at H, and M is the midpoint of BC. The circumcircle of $\triangle ADE$ intersects the circumcircle of $\triangle ABC$ at points K and A. Prove that points M, H, and K are collinear.

Proof. Let the circumcentre of $\triangle ABC$ be the origin, then $H = A + B + C$, and we have $2(A - K)\left(A + B + C - \frac{B+C}{2}\right) + (A - K)(K - (A + B + C)) - (A^2 - K^2) = 0$.

Figure 3.37

80   Solving Problems in Point Geometry

Note. First, the points A, D, E, H, and K are concyclic, with AH as the diameter. Therefore, the original proposition is equivalent to proving that $AK \perp HM$. Additionally, all the identities in this problem involve $A - K$, making it more convenient for verification, reminiscent of the distributive property in algebra, where solving geometric problems is akin to solving algebraic ones.

Example 17. In Figure 3.38, in $\triangle ABC$, where $AB = AC$, CD is perpendicular to CA, E is on BC, and $DC = DE$. Prove that $AB \perp DE$.

Proof. Let $A = 0$, then we have

$$B(D-E) + \left(D - \frac{E+C}{2}\right)(C-B) - \frac{B+C}{2}(C-E) + C(C-D) = 0.$$

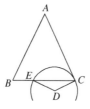   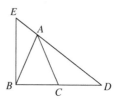

Figure 3.38   Figure 3.39

Example 18. In Figure 3.39, the side BC of $\triangle ABC$ is extended to point D such that $CD = BC$. A perpendicular line is drawn from point B on line BC to the extension of DA at point E. If $AD = 3AE$, what type of triangle is $\triangle ABC$?

Proof. Let $A = 0$, then we have $(B^2 - C^2) + \frac{3}{2}(B - C)\left(-\frac{2C-B}{3} - B\right) = 0$. This implies that $AB = AC$, so $\triangle ABC$ is an isosceles triangle.

Example 19. In Figure 3.40, points D and E are located on the sides AB and BC of equilateral $\triangle ABC$, respectively, such that $3AD = AB$ and $3BE = BC$. Lines AE and CD intersect at point P. Prove that $PB \perp PC$.

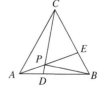

Figure 3.40

Proof 1. $\left(\frac{4A+2B+C}{7} - B\right)\left(\frac{4A+2B+C}{7} - C\right) - \frac{6}{49}\left[(A-B)^2 - (A-C)^2\right] + \frac{16}{49}\left[(C-B)^2 - (A-C)^2\right] = 0.$

Proof 2. Let $B = 0$ and $|AB| = 1$. Then,

$$\left(\frac{4A+2B+C}{7} - B\right)\left(\frac{4A+2B+C}{7} - C\right)$$
$$= \frac{4A+C}{7} \cdot \frac{4A-6C}{7} = \frac{1}{49}(16 - 6 - 12 + 2) = 0.$$

Note. The key to this problem lies in quickly determining $P = \frac{4A+2B+C}{7}$. In a formal examination, it is important to provide a detailed explanation of this step. The advantage of using identities is that they can lead to the discovery of new propositions, although they may or may not represent the quickest approach to solving the original problem.

Example 20. In Figure 3.41, point C is located on line segment AB. On the same side of segment AB, construct squares $ACDE$ and $CBFG$. Prove that $AG \perp DB$.

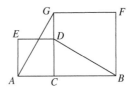

Figure 3.41

Proof.

$$(B-D)(A-G) + (C-B)(C-G)$$
$$+ (C-A)(C-D) + [(A-C)(C-B)$$
$$- (C-D)(C-G)] = 0.$$

Note. The term $(A-C)(C-B) - (C-D)(C-G) = 0$ is often overlooked. From the identity, it can be observed that the original problem can be extended to the following: given point C on line segment AB, construct rectangles $ACDE$ and $CBFG$ on the same side of segment AB, such that $\frac{AC}{CD} = \frac{CG}{CB}$. Prove that $AG \perp DB$.

Example 21. In Figure 3.42, in scalene acute $\triangle ABC$ with centroid G, let M be the midpoint of BC. A circle with centre G and radius GM intersects BC at point N, other than M. If S is the point symmetric to A about N, prove that $GS \perp BC$. (the 2018 Italian Mathematical Olympiad)

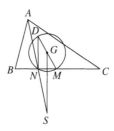

Figure 3.42

Proof. Let $S = 2N - A$ and $O = 2\frac{A+B+C}{3} - \frac{B+C}{2}$:

$$\left(2N - A - \frac{A+B+C}{3}\right)(B - C)$$
$$- 2\left[N - \left(2\frac{A+B+C}{3} - \frac{B+C}{2}\right)\right](B - C) = 0.$$

Note. By drawing the diameter MD, we cleverly constrain N to lie on BC. In essence, it is a transformation of the equality $2N - A - \frac{A+B+C}{3} = 2\left[N - \left(2\frac{A+B+C}{3} - \frac{B+C}{2}\right)\right]$.

Example 22. In Figure 3.43, in acute $\triangle ABC$, AH is an altitude and AM is a median, and points X and Y are on lines AB and AC, respectively, such that $AX = XC$ and $AY = YB$. N is the midpoint of XY. Prove that $NM = NH$. (2018 Ukrainian Mathematical Olympiad)

Proof.

$$\left(\frac{X+Y}{2} - \frac{\frac{B+C}{2} + H}{2}\right)(B - C) + \frac{1}{2}(B - C)(H - A)$$
$$+ \left(X - \frac{A+C}{2}\right)\left(\frac{A+C}{2} - Y\right)$$
$$- \left(Y - \frac{A+B}{2}\right)\left(\frac{A+B}{2} - X\right) = 0.$$

From the given identity, we can deduce a more general proposition: as shown in Figure 3.44, in $\triangle ABC$, AH is perpendicular to BC, AM is the median, points X and Y lie on lines AB and AC, respectively, with $AX = XC$ and $AY = YB$, N is the midpoint of XY, and K is the midpoint of MH. Prove that $NK \perp BC$.

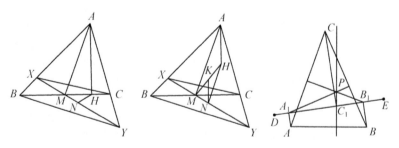

Figure 3.43   Figure 3.44   Figure 3.45

Example 23. In Figure 3.45, on the plane of $\triangle ABC$, there is a line DE. From point A, draw $A_1A \perp DE$ to point A_1; from point B, draw $B_1B \perp DE$ to point B_1; and from point C, draw $C_1C \perp DE$ to point C_1. Draw A_1P perpendicular to BC from A_1, and draw B_1P perpendicular to CA from B_1. Prove that $C_1P \perp AB$.

Proof. Let $P = 0$, then

$$C_1(A - B) + A_1(B - C) + B_1(C - A) - (B_1 - C_1)(A_1 - A)$$
$$- (C_1 - A_1)(B_1 - B) - (A_1 - B_1)(C_1 - C) = 0.$$

Example 24. In Figure 3.46, point M inside $\triangle ABC$ is such that $AM = \frac{AB}{2}$ and $CM = \frac{BC}{2}$. Points C_0 and A_0 lying on AB and CB, respectively, are such that $BC_0 : AC_0 = BA_0 : CA_0 = 3$. Prove that $MC_0 = MA_0$. (2019 Sarykin Geometry Olympiad)

Proof. Let $M = 0$, then

$$16\left[\left(\frac{3A+B}{4}\right)^2 - \left(\frac{3C+B}{4}\right)^2\right]$$
$$- 3(4A^2 - (A-B)^2) + 3(4C^2 - (C-B)^2) = 0.$$

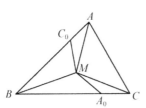 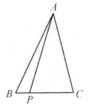

Figure 3.46   Figure 3.47

Example 25 (Stewart's theorem). As shown in Figure 3.47, let P be a point on the side BC of $\triangle ABC$. Then, we have

$$AB^2 \cdot PC + AC^2 \cdot PB = AP^2 \cdot BC + BP \cdot PC \cdot BC.$$

Proof. We rewrite it as follows:

$$AP^2 = AB^2 \cdot \frac{PC}{BC} + AC^2 \cdot \frac{PB}{BC} - BC^2 \cdot \frac{BP}{BC} \cdot \frac{PC}{BC}.$$

If $\frac{BP}{BC} = k$, then $AP^2 = (1-k)AB^2 + kAC^2 - k(1-k)BC^2$. Now, let $A = 0$ and $P = kC + (1-k)B$. Then,

$$(1-k)(A-B)^2 + k(A-C)^2 - k(1-k)(B-C)^2 - (A-P)^2 = 0.$$

Note. The statement implies that any problem solvable using Stewart's theorem can be represented as an identity. The following special cases are commonly used:

- If $AB = AC$, then $AP^2 = AB^2 - BP \cdot PC$.
- If $BP = PC$, then $AP^2 = \frac{1}{2}AB^2 + \frac{1}{2}AC^2 - \frac{1}{4}BC^2$.
- If $\angle BAP = \angle CAP$, then $AP^2 = AB \cdot AC - BP \cdot PC$.

Example 26. In Figure 3.48, given trapezium $ABCD$ with $AD \parallel BC$ and $\angle A = 90°$, E is the midpoint of AB, and $DE \perp CE$, prove that $AB^2 = 4AD \cdot BC$.

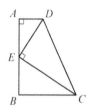

Figure 3.48

Proof.

$$[(A-B)^2 - 4(A-D)(B-C)] - 2(A-D)(A-B)$$
$$- 2(B-A)(B-C) + 4\left(\frac{A+B}{2} - C\right)\left(\frac{A+B}{2} - D\right) = 0.$$

Note. From geometry problems, we can generate identities, thereby proving the original proposition and discovering new ones. Can we create identities and generate new propositions out of thin air? Of course, we can. Generating algebraic identities is often easier than studying geometric figures. Therefore, this is another approach to creating propositions. Readers interested in exploring propositions can practice this method on their own. For instance, following this example, you can derive more identities:

$$(A-B)^2 - \frac{9}{2}(A-D)(B-C) - 3(A-D)(A-B)$$
$$-\frac{3}{2}(B-A)(B-C) + \frac{9}{2}\left(\frac{2A+B}{3} - C\right)\left(\frac{2A+B}{3} - D\right) = 0,$$
$$\left[(A-B)^2 - \frac{16}{3}(A-D)(B-C)\right] - 4(A-D)(A-B)$$
$$-\frac{4}{3}(B-A)(B-C) + \frac{16}{3}\left(\frac{3A+B}{4} - C\right)\left(\frac{3A+B}{4} - D\right) = 0.$$

The method of identities has several advantages:

1. It transforms geometric proofs into algebraic calculations, making the process more straightforward. It builds a bridge between algebraic and geometric properties, making the connection between the validity of geometric properties and the validity of algebraic expressions much stronger.
2. It offers concise representations. A single equation completes the proof, and the expression is often shorter and more straightforward than the original problem. The provided identity-based proofs usually require simple calculations for verification and carry a clear geometric meaning, making them easily understandable to readers without the need for intricate deductive reasoning.
3. It allows for equivalent reasoning about necessary and sufficient conditions in geometry. This method deepens the understanding of relationships between conditions, helps in finding redundant conditions, and generates new propositions. It provides rich material for exploring variations on a single problem.

Identities may appear as just a single line, but they contain a wealth of information and can be used to create new problems. Readers interested in solving problems using identities are encouraged to try creating identities themselves, especially if they have access to a computer. To facilitate better communication with others, it is also important to master the conversion between point geometry identity methods and general vector methods.

Exercise 3.3

1. In Figure 3.49, in trapezium $ABCD$, where $AD \parallel BC$ and $AB \perp BC$, prove that $AC \perp BD \iff AB^2 = BC \cdot AD$.

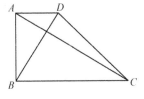

Figure 3.49   Figure 3.50

2. In Figure 3.50, in isosceles $\triangle ABC$, where $AB = AC$, a line CD is drawn perpendicular to AB at point D. Prove that $BC^2 = 2BD \cdot BA$.

3. In Figure 3.51, in $\triangle ABC$, where $AB = AC$, a point D is taken on the extension of BA, and a point E is taken on AC such that $AE = AD$. Connect DE. Prove that $DE \perp BC$.

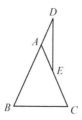

Figure 3.51

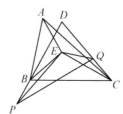

Figure 3.52

4. In Figure 3.52, in $\triangle ABC$, E is the orthocentre of $\triangle DBC$. Points P and Q are on lines DB and DC, respectively, such that $PE \perp AC$ and $QE \perp AB$. Prove that $AE \perp PQ$.
5. H, G, and O are the orthocentre, centroid, and circumcentre of $\triangle ABC$, respectively. M is the midpoint of segment HG. Prove that $MA^2 + MB^2 + MC^2 = 3OA^2$.
6. In Figure 3.53, points C and D on different sides of AB are on the circumference of the circle such that the extensions of chords CA and BD intersect at point E. Point F is a point on the line AB such that $BA \cdot BF = BD \cdot BE$. Prove that $EF \perp AB$.

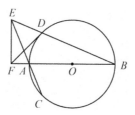

Figure 3.53

7. In Figure 3.54, in $\triangle ABC$, $CA \perp CB$, and $BC = 2AC$. D is the midpoint of BC, and E is a point on line AD such that $AE : ED = 1 : 2$. Prove that $CE \perp AB$.
8. In Figure 3.55, $\triangle ABC$ is an isosceles right-angled triangle with $\angle C = 90°$, and points M and N are the midpoints of AC and BC, respectively. Point D lies on ray BM such that $BD = 2BM$, and point E

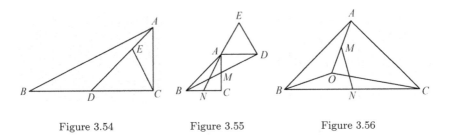

Figure 3.54   Figure 3.55   Figure 3.56

lies on ray NA such that $NE = 2NA$. Prove that $BD \perp DE$. (1996 Tianjin Competition)

9. In Figure 3.56, in $\triangle ABC$ with $AB \perp AC$ and for any point O, let M and N be the midpoints of segments OA and BC, respectively. Prove that $MN = \frac{1}{2}\sqrt{OB^2 + OC^2 - OA^2}$.

10. In Figure 3.57, H is the orthocentre of $\triangle ABC$, and D is symmetric to H about line BC. The midpoint of AH is K. Prove that points B, D, E, and K are concyclic.

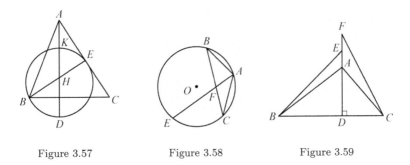

Figure 3.57   Figure 3.58   Figure 3.59

11. In Figure 3.58, in circle O, $AB = AC$, and chord AE intersects BC at F. Prove that $AB^2 = AE \cdot AF$.
12. In Figure 3.59, AD is an altitude of $\triangle ABC$, and points E and F lie on the extension of the line AD, such that $DE = AC$ and $DF = AB$. Prove that $BE = CF$.
13. In Figure 3.60, in trapezium $ABCD$, $AD \parallel BC$ and $\angle ABC = 90°$, and the diagonal BD is perpendicular to the side DC. Prove that $BD^2 = AD \cdot BC$ (1992 Shanghai Secondary School Entrance Examination).

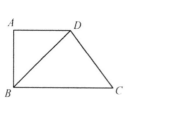
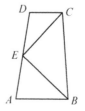

Figure 3.60 Figure 3.61

14. In Figure 3.61, in quadrilateral $ABCD$, $AB \parallel DC$, and point E is the midpoint of AD. Prove that $EC \perp EB \iff AB + DC = BC$.
15. Given trapezium $ABCD$ with $AB \parallel DC$, prove that $AC^2 + BD^2 = (AB + CD)^2 \iff AC \perp BD$.

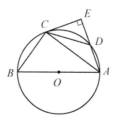

Figure 3.62

16. In Figure 3.62, $\triangle ABC$ is inscribed in circle O with AB as a diameter, and point D lies on the circle O. The line tangents to the circle O at point C and intersects the extension of AD at point E, and $AE \perp CE$. Prove that $DC = BC$.
17. In Figure 3.63, in convex quadrilateral $ABCD$, the diagonals AC and BD intersect at point O, and M and N are the midpoints of sides AB and DC, respectively. Prove that $AC^2 + BD^2 = 4MN^2$.
18. In Figure 3.64, in $\triangle ABC$, $\angle C = 90°$, M is the midpoint of side BC, and $MD \perp AB$ at point D. Prove that $AD^2 = AC^2 + BD^2$.

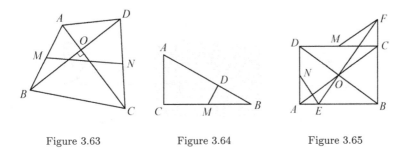

Figure 3.63 Figure 3.64 Figure 3.65

19. In Figure 3.65, let O be the centre of rectangle $ABCD$ ($AB \neq BC$). The perpendicular from O to BD intersects AB and BC at points E and F, respectively. M and N are the midpoints of CD and AD, respectively. Prove that $FM \perp EN$. (2008 Romania Mathematics Olympiad)
20. In Figure 3.66, $\triangle ABC$ is an acute triangle, H is the orthocentre, and M is the midpoint of AH. Line $MK \parallel AB$ is drawn through M, where K is on BC. Line $KN \perp BC$ is drawn through K, where N is on AB. Prove that $NM \perp CM$.

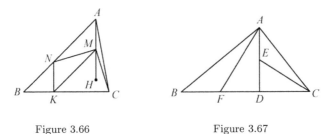

Figure 3.66 Figure 3.67

21. In Figure 3.67, AD is an altitude of $\triangle ABC$. E is the midpoint of AD, and F is the midpoint of BD. Prove that $AF \perp CE \iff AB \perp AC$.
22. In Figure 3.68, $\triangle ABC$ is an equilateral triangle, D is a point on side AC, and E is a point on the extension of side BC, such that $CE = AD$. Prove that $DB = DE$.

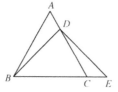
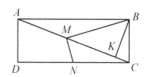

Figure 3.68   Figure 3.69

23. In Figure 3.69, in rectangle $ABCD$, $BK \perp AC$ at point K, and M and N are the midpoints of AK and DC, respectively. Prove that $MB \perp MN$.
24. In Figure 3.70, H is the orthocentre of acute $\triangle ABC$, and a circle with diameter AH intersects the circumcircle of $\triangle BHC$ at points H and P. Prove that point P lies on the median of side BC.
25. In Figure 3.71, in $\triangle ABC$, point D is on AB such that $3AD = AB$, and $BE \perp CD$. Point F is the midpoint of CE. Prove that $AB = AC \iff FA \perp FB$.

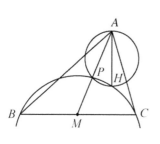
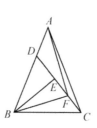
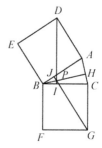

Figure 3.70   Figure 3.71   Figure 3.72

26. In Figure 3.72, in $\triangle ABC$, squares $BADE$ and $CBFG$ are constructed externally. I is the foot of the perpendicular from D to BC, and J is the foot of the perpendicular from G to BA. The two perpendiculars intersect at point P. Prove that $BP \perp AC$.
27. In Figure 3.73, consider the right triangles OAB and ODC sharing a common vertex. Given $OC = OB$ and $DB \perp AO$, prove that $AC \perp OD$.

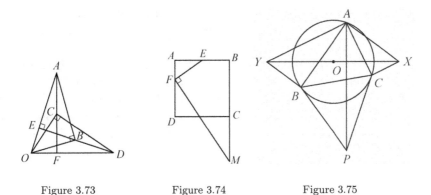

Figure 3.73 Figure 3.74 Figure 3.75

28. In Figure 3.74, in square $ABCD$, E is the midpoint of AB, and F is a point on AD such that $2AF = FD$. M is a point on the extension of BC such that $6BM = 11BC$. Prove that $FE \perp FM$.

29. In Figure 3.75, let $\triangle ABC$ be an acute triangle, and let P be the intersection of the tangents at B and C of the circumscribed circle of $\triangle ABC$. The line through A perpendicular to AB intersects with the line through C perpendicular to AC at X, and the line through A perpendicular to AC intersects with the line through B perpendicular to AB at Y. Show that $AP \perp XY$. (2020 Dutch IMO Team Selection Exam)

30. In Figure 3.76, consider $\triangle ABC$, where $AB \perp AC$. Let D be a point on side BC. $DE \perp AB$, $DF \perp AC$, and $AG \perp BC$. Prove that $EG \perp FG$.

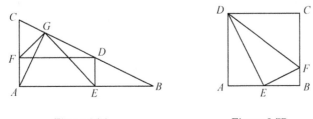

Figure 3.76 Figure 3.77

31. In Figure 3.77, in square $ABCD$, E is the midpoint of AB, and $3BF = FC$. Prove that $DE \perp EF$.

32. Given that G is the centroid of quadrilateral $ABCD$, and M and N are the midpoints of AD and BC, respectively, prove that $GA \perp GD \iff AD = MN$.
33. In Figure 3.78, AC intersects BD at P such that $PA = PD$ and $PB = PC$, and O is the circumcentre of $\triangle PAB$. Prove that $OP \perp CD$.
34. In Figure 3.79, consider quadrilateral $ABCD$ with diagonals intersecting at point P. Let Q be a point on the perpendicular bisector of segment PD such that $QP \perp AB$. Prove that Q is the circumcentre of $\triangle CDP$.

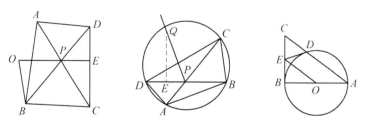

Figure 3.78   Figure 3.79   Figure 3.80

35. In Figure 3.80, in $\triangle ABC$, where O is the midpoint of AB, D is a point on AC, and E is the midpoint of BC. Prove that, given any two of the conditions $BA \perp BC$, $DE \perp DO$, and $DB \perp DA$, the remaining condition holds.
36. In Figure 3.81, in quadrilateral $ABCD$, where $AC \perp BD$, E and F be the midpoints of AB and AD, respectively. Lines through E and F are perpendicular to CD and BC, respectively, and they intersect at point G. Prove that $CG \perp BD$.

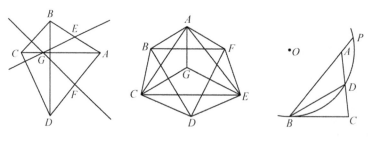

Figure 3.81   Figure 3.82   Figure 3.83

37. In Figure 3.82, given that $AG \perp BF$, $CG \perp BD$, and $EG \perp DF$, prove that $AB^2 + CD^2 + EF^2 = BC^2 + DE^2 + FA^2$.
38. In Figure 3.83, in $\triangle ABC$, D is the midpoint of side AC. A circle passing through D and tangent to side BC at point B intersects line AB at point P. Prove that $BA \cdot BP = 2BD^2$. (2017–2018 British Mathematical Olympiad Round 2)
39. In Figure 3.84, ω denotes the circumcircle of $\triangle ABC$. A tangent line to ω from point A intersects line BC at point D. M is the midpoint of BC. The line AM intersects ω at the other point N. Point K is chosen such that quadrilateral $ADMK$ is a parallelogram. Prove that $KA = KN$. (2019 International Mathematical Excellence Olympiad, IMEO)

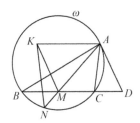
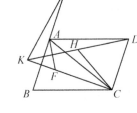

Figure 3.84   Figure 3.85

40. Let G be the centroid and O be the circumcentre of $\triangle ABC$. If the line connecting G and O is perpendicular to the median AM, prove that AB^2, BC^2, CA^2 form an arithmetic progression.
41. In Figure 3.85, consider parallelogram $ABCD$. Extend line BA to point E such that $AB = AE$. If there exists a point K such that H is the midpoint of KD, F is the trisection point of KC (closer to point K), $EK = AC$, and $AB = CH$, prove that $AF \perp KD$.

Chapter 4
Identity-Based Method 2: Undetermined Coefficients

4.1 Method of Undetermined Coefficients

In the previous chapter, we used the method of analysis to establish identities. This method involves examining the terms present in the conclusion and finding ways to eliminate them. During this process, new terms might be generated, and again, we try to eliminate them. This continues until we arrive at the desired identity. This approach works for relatively simple problems, but it becomes challenging when dealing with many conditions. In this chapter, we introduce a somewhat "clumsy" method, which involves expressing the conclusion and known conditions in geometric terms as much as possible and then using the method of undetermined coefficients to see if we can establish an identity. Of course, the best approach is a combination of both methods: thinking about how terms can cancel each other out and using brute-force calculations to balance coefficients.

To keep the solving process concise, the first step is to find a relatively simple way to represent geometric relationships, such as midpoints, perpendicularity, equality of line segments, and concyclic conditions. Next, we connect the given conditions and conclusions (which are essentially various geometric relationships) in a straightforward manner. We find that using point geometry representations is concise, and it is possible to string conditions together using the concept of linear dependence.

Example 1. In Figure 4.1, AD and BE are altitudes of $\triangle ABC$, and F and G are midpoints of AB and DE, respectively. Prove that $FG \perp ED$.

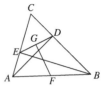

Figure 4.1

Proof. Traditional approach:

$$\left.\begin{array}{l}\left.\begin{array}{l}AD \perp BC \Rightarrow \text{Rt } \triangle ADB \\ F \text{ is the midpoint of } AB\end{array}\right\} \Rightarrow AF = DF \\ \left.\begin{array}{l}EB \perp AC \Rightarrow \text{Rt } \triangle AEB \\ F \text{ is the midpoint of } AB\end{array}\right\} \Rightarrow AF = EF\end{array}\right\} \Rightarrow \left.\begin{array}{l}EF = DF \\ \\ G \text{ is the midpoint of } DE\end{array}\right\}$$

$\Rightarrow ED \perp FG$.

The traditional approach to problem-solving encounters several difficulties. First, it is often challenging to determine which auxiliary lines to draw. For example, in Example 1, the lines EF and DF have multiple roles: they are both medians of the right-angled $\triangle EAB$ and two equal length sides of the isosceles $\triangle FDE$. Second, applying the known conditions can be tricky, as it involves deciding when to use them, how to combine them with other conditions, and which rules of inference to apply. In Example 1, the points F and G are both midpoints, but they are used differently. Traditional reasoning often follows a step-by-step deductive pattern, and if any step is not clarified, the entire reasoning process may fail.

The addition of auxiliary lines follows certain patterns, but in many cases, it requires creative insights. Therefore, we aim to introduce a new method of proof that avoids the need for auxiliary lines. Simultaneously, we aim to shorten the reasoning process by replacing logical deductions with computations.

Analysis.

(1) The problem involves seven points: A, B, C, D, E, F, and G. Based on geometric relations, we have $F = \frac{(A+B)}{2}$ and $G = \frac{(D+E)}{2}$. This reduces the number of variables.

(2) The essence of Example 1 is how to derive the conclusion polynomial $(F-G)(D-E) = 0$ from the condition polynomials $(A-E)(B-E) = 0$ and $(A-D)(B-D) = 0$.

(3) Express the conclusion expression as a linear combination of conditional expressions F:

$$\left(\frac{A+B}{2} - \frac{D+E}{2}\right)(D-E) + k_1(A-E)(B-E)$$
$$+ k_2(A-D)(B-D) = 0.$$

(4) Expand F into an equation with the basic points A, B, C, D, and E as variables:

$$\left(\frac{1}{2} + k_1\right) E^2 + \left(-\frac{1}{2} - k_1\right) BE + \left(\frac{1}{2} - k_2\right) BD + \left(-\frac{1}{2} + k_2\right) D^2$$
$$+ \left(-\frac{1}{2} - k_1\right) AE + \left(\frac{1}{2} - k_2\right) AD + (k_1 + k_2) AB = 0.$$

(5) Solve the system of equations:

$$\frac{1}{2} + k_1 = -\frac{1}{2} - k_1 = \frac{1}{2} - k_2 = -\frac{1}{2} + k_2$$
$$= -\frac{1}{2} - k_1 = \frac{1}{2} - k_2 = k_1 + k_2 = 0,$$

which yields $k_1 = -\frac{1}{2}$ and $k_2 = \frac{1}{2}$. Thus, we obtain the identity

$$\left(\frac{A+B}{2} - \frac{D+E}{2}\right)(D-E) - \frac{1}{2}(A-E)(B-E)$$
$$+ \frac{1}{2}(A-D)(B-D) = 0.$$

From this identity, and using $(A-E)(B-E) = 0$ and $(A-D)(B-D) = 0$, we can deduce

$$\left(\frac{A+B}{2} - \frac{D+E}{2}\right)(D-E) = 0,$$

which implies $FG \perp ED$. We note that, for the conclusion to hold, point C plays no role and is redundant, and the four points A, B, D, and E need not necessarily lie on the same plane. Such insights deepen our understanding of the problem and can be extended to higher dimensions.

n polynomials add up to zero, where $n-1$ of them are equal to zero, and the remaining one is naturally zero. This seemingly trivial principle has clever applications. It can transform Example 1 into five different problems (sometimes, the idea of the identical method is needed):

(1) In the spatial quadrilateral $ABDE$, where F is the midpoint of AB and G is the midpoint of DE, if $EA \perp EB$ and $DA \perp DB$, then $FG \perp ED$.
(2) In the spatial quadrilateral $ABDE$, where F is the midpoint of AB and G is the midpoint of DE, if $EA \perp EB$ and $FG \perp ED$, then $DA \perp DB$.
(3) In the spatial quadrilateral $ABDE$, where F is the midpoint of AB and G is the midpoint of DE, if $FG \perp ED$ and $DA \perp DB$, then $EA \perp EB$.
(4) In the spatial quadrilateral $ABDE$, where F is the midpoint of AB and G is a point on DE, if $FG \perp ED$, $EA \perp EB$, and $DA \perp DB$, then G is the midpoint of DE.
(5) In the spatial quadrilateral $ABDE$, where G is the midpoint of DE and F is a point on AB, if $FG \perp ED$, $EA \perp EB$, and $DA \perp DB$, then F is the midpoint of AB.

Example 2. In $\triangle ABC$, as shown in Figure 4.2, extend BC to D such that $CD = BC$, and extend CA to E such that $AE = 2CA$. If $\angle BAC = 90°$, prove that $AD = BE$.

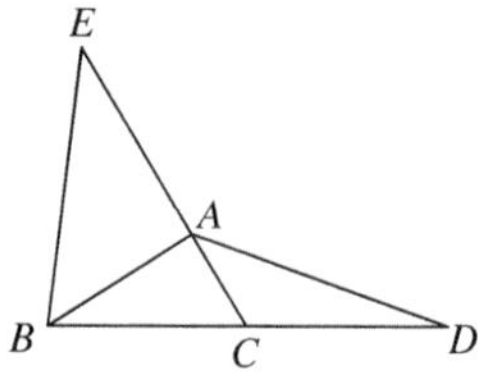

Proof. $(A - (2C - B))^2 - (B - (3A - 2C))^2 + 8(A - B)(A - C) = 0$.

Analysis.

Figure 4.2

(1) The problem involves five points: A, B, C, D, and E. Based on geometric relationships, we have $D = 2C - B$ (equivalent to $C = \frac{D+B}{2}$) and $E = 3A - 2C$, which reduces the number of variables.
(2) Using the perpendicular relationship $AB \perp AC$, we can write the conditional polynomial as $(A - B)(A - C) = 0$. Based on the equality $AD = BE$, we can write the conclusion polynomial as $(A - (2C - B))^2 - (B - (3A - 2C))^2 = 0$.
(3) Express the conclusion as a linear combination of conditional expressions, denoted as F:

$$(A - (2C - B))^2 - (B - (3A - 2C))^2 + k_1(A - B)(A - C) = 0.$$

(4) Expand F as an equation in terms of the variables A, B, and C:

$$(-8 + k_1)A^2 + (8 - k_1)AB + (8 - k_1)AC + (-8 + k_1)BC = 0.$$

(5) Solve the coefficient equations: $-8 + k_1 = 8 - k_1 = 8 - k_1 = -8 + k_1 = 0$. Solving for k_1, we find $k_1 = 8$, which leads to the identity

$$(A - (2C - B))^2 - (B - (3A - 2C))^2 + 8(A - B)(A - C) = 0.$$

This problem has relatively simple conditions, and it can also be solved by directly observing the coefficient of A^2 to obtain $k_1 = 8$ without using the method of undetermined coefficients. For solving fill-in-the-blank or multiple-choice questions, the observation method is often quicker, as seen in Example 7.

Let $A = 0$ and $E = -2C$, then the identity becomes simpler:

$$(2C - B)^2 - (B - (-2C))^2 + 8BC = 0.$$

After obtaining the identity $F = 0$, you can substitute the known conditions into the identity, making several terms zero, which immediately leads to the conclusion. From the identity and $(A - B)(A - C) = 0$, we can deduce the following:

$$(A - (2C - B))^2 - (B - (3A - 2C))^2 = 0,$$

which implies $|AD| = |BE|$. In mathematics, it is often necessary to study inverse propositions. In this problem, the identity easily yields

$$|AD| = |BE| \iff \angle BAC = 90°.$$

When the sum of n items equates to 0, where $n - 1$ of the items are 0, it can be inferred that the rest item is also 0. This simple principle significantly enhances the value of an identity. Upon establishing an identity, numerous new propositions can be derived from the original one. This aspect is tremendously valuable in the exploration of variations in geometric problems, which, due to limited space, will not be further delineated.

A one-line proof, as referred to in geometric competition problems, means expressing the entire proof in one line, or, more precisely, using a single identity. However, obtaining this one-line proof may require some effort, similar to seeing someone use the SOS method for inequalities without knowing the effort behind it. So-called "instant kills" are often just for show.

Consider that in teaching or examinations, the representation in terms of point vectors might not always be acceptable. After obtaining the identity, it can be rewritten in a general vector form.

Note. Let $A = 0$, and all vectors should be rewritten as vectors starting with A. The approach for rewriting is as follows:

$AB \perp AC$ (i.e., $(A - B)(A - C) = 0$) can be rewritten as

$$\overrightarrow{AB} \cdot \overrightarrow{AC} = 0. \tag{1}$$

$AD = BE$ (i.e., $(A - (2C - B))^2 - (B - (3A - 2C))^2 = 0$) can be rewritten as

$$(\overrightarrow{AA} - (2\overrightarrow{AC} - \overrightarrow{AB}))^2 - (\overrightarrow{AB} - (3\overrightarrow{AA} - 2\overrightarrow{AC}))^2 = 0,$$

which simplifies to

$$\overrightarrow{AD}^2 - \overrightarrow{BE}^2 = (2\overrightarrow{AC} - \overrightarrow{AB})^2 - (\overrightarrow{AB} - (-2\overrightarrow{AC}))^2 = -8\overrightarrow{AB} \cdot \overrightarrow{AC} = 0. \tag{2}$$

It is easy to see that $8 \times (1) + (2) = 0$, which completes the proof.

It is easy to observe that the method of identities serves as a concise representation and comprehensive approach, akin to the general vector method. These two methods can be interchanged, with the emphasis placed on whether the conclusion polynomial can be expressed in terms of the condition polynomials, rather than scrutinizing the individual expressions of each term.

Example 3. In Figure 4.3, within parallelogram $ABCD$, where M is the midpoint of BC, and P is any point (not necessarily on the plane of ABC), we aim to prove three conditions: $PA = PD$, $CP \perp AB$, and $MA \perp MP$. If any two of these conditions are known, the rest can be deduced.

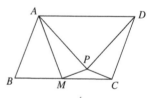

Figure 4.3

We start the proof with the following identity:

$$(P - A)^2 - (P - (A + C - B))^2 + 4(C - P)(A - B)$$
$$+ 4\left(\frac{B+C}{2} - A\right)\left(\frac{B+C}{2} - P\right) = 0.$$

It is worth noting that we let $P = 0$, and all vectors should be rewritten as vectors starting with P:

$$[A^2 - (A + C - B)^2] + 4C(A - B) + 4\frac{B+C}{2}\left(\frac{B+C}{2} - A\right) = 0.$$

Rewrite as follows:

$PA = PD$ is equivalent to $\overrightarrow{PA}^2 - (\overrightarrow{PA} + 2\overrightarrow{PC} - \overrightarrow{PB})^2 = 0$, that is,

$$-\overrightarrow{PC}^2 - \overrightarrow{PB}^2 - 2\overrightarrow{PA} \cdot \overrightarrow{PC} + 2\overrightarrow{PA} \cdot \overrightarrow{PB} + 2\overrightarrow{PB} \cdot \overrightarrow{PC} = 0. \quad (1)$$

Similarly, $CP \perp AB$ is equivalent to

$$\overrightarrow{PC} \cdot \overrightarrow{BA} = 0, \text{ that is,}$$

$$\overrightarrow{PC} \cdot \overrightarrow{PA} - \overrightarrow{PC} \cdot \overrightarrow{PB} = 0. \quad (2)$$

$MA \perp MP$ is equivalent to $\overrightarrow{AM} \cdot \overrightarrow{PM} = 0$, i.e., $\left(\frac{\overrightarrow{PB}+\overrightarrow{PC}}{2} - \overrightarrow{PA}\right) \cdot \left(\frac{\overrightarrow{PB}+\overrightarrow{PC}}{2}\right) = 0$, that is,

$$\frac{\overrightarrow{PB}^2 + \overrightarrow{PC}^2 + 2\overrightarrow{PB} \cdot \overrightarrow{PC} - 2\overrightarrow{PA} \cdot \overrightarrow{PB} - 2\overrightarrow{PA} \cdot \overrightarrow{PC}}{4} = 0. \quad (3)$$

It is evident that combining equations (1), (2), and (3), we obtain the desired result: $(1) + 4 \times (2) + 4 \times (3) = 0$. Thus, we simultaneously prove all three propositions.

Example 4. In Figure 4.4, in $\triangle ABC$, where D is the midpoint of AB, and E and F are points on lines CA and CB, respectively, prove the following:

1. $EF^2 = EA^2 + BF^2$,
2. $DE \perp DF$,
3. $CA \perp CB$.

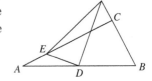

Figure 4.4

Any two of these conditions imply the rest.

Proof.

$$2(A-E)(B-F) + [(E-A)^2 + (B-F)^2 - (E-F)^2]$$
$$-4\left(\frac{A+B}{2} - E\right)\left(\frac{A+B}{2} - F\right) = 0.$$

On rewriting,

1. $EF^2 = EA^2 + BF^2$ is equivalent to $(\overrightarrow{CF} - \overrightarrow{CE})^2 - (\overrightarrow{CA} - \overrightarrow{CE})^2 - (\overrightarrow{CF} - \overrightarrow{CB})^2 = 0$, i.e.,

$$-\overrightarrow{CA}^2 - \overrightarrow{CB}^2 - 2\overrightarrow{CE} \cdot \overrightarrow{CF} + 2\overrightarrow{CA} \cdot \overrightarrow{CE} + 2\overrightarrow{CB} \cdot \overrightarrow{CF} = 0. \quad (1)$$

2. $DE \perp DF$ is equivalent to $(\vec{CE} - \vec{CD}) \cdot (\vec{CF} - \vec{CD}) = 0$, i,e.,

$$\left(\vec{CE} - \frac{\vec{CA} + \vec{CB}}{2}\right) \cdot \left(\vec{CF} - \frac{\vec{CA} + \vec{CB}}{2}\right) = 0,$$

$$4\vec{CE} \cdot \vec{CF} - 2\vec{CE} \cdot \vec{CA} - 2\vec{CE} \cdot \vec{CB}$$
$$- 2\vec{CF} \cdot \vec{CA} - 2\vec{CF} \cdot \vec{CB} + 2\vec{CA} \cdot \vec{CB}$$
$$+ \vec{CA}^2 + \vec{CB}^2 = 0. \qquad (2)$$

3. $CA \perp CB$ is equivalent to $EA \perp FB$, $(\vec{CA} - \vec{CE}) \cdot (\vec{CB} - \vec{CF}) = 0$, i.e.,

$$\vec{CA} \cdot \vec{CB} - \vec{CA} \cdot \vec{CF} - \vec{CB} \cdot \vec{CE} + \vec{CE} \cdot \vec{CF} = 0. \qquad (3)$$

Adding equations (1) and (2) while subtracting twice equation (3) results in 0, thereby proving all three propositions simultaneously.

Example 5. In Figure 4.5, quadrilateral $ABCD$ is inscribed by a circle, where diagonals AC and BD intersect at point P, and point Q lies on line BD (other than B) such that $PQ = PB$. On the plane, there are two points R and S such that quadrilaterals $CAQR$ and $DBCS$ are parallelograms. Prove that points C, Q, R, and S are concyclic. (2011 Macedonian National Team Selection Exam)

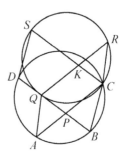

Figure 4.5

Proof. Let $P = 0$, $Q = -B$, $S = C + D - B$, $R = Q + C - A$, and $K = Q + C - P$. We have

$$[(K - C)(K - S) - (K - Q)(K - R)] + (CA - DB) = 0.$$

From this identity, we can see that points A, B, C, and D are concyclic if and only if points C, Q, R, and S are concyclic.

Rewrite the condition that A, B, C, and D are concyclic as $\vec{PA} \cdot \vec{PC} - \vec{PB} \cdot \vec{PD} = 0$.

Now, let CS intersect QR at point K. The condition that points Q, C, R, and S are concyclic can be rewritten as

$$\vec{KS} \cdot \vec{KC} - \vec{KR} \cdot \vec{KQ} = 0,$$

which is equivalent to
$$(\overrightarrow{PS} - \overrightarrow{PK}) \cdot (\overrightarrow{PC} - \overrightarrow{PK}) - (\overrightarrow{PR} - \overrightarrow{PK}) \cdot (\overrightarrow{PQ} - \overrightarrow{PK}) = 0,$$
i.e.,
$$(\overrightarrow{PC} + \overrightarrow{PD} - \overrightarrow{PB} - (\overrightarrow{PC} - \overrightarrow{PB})) \cdot (\overrightarrow{PC} - (\overrightarrow{PC} - \overrightarrow{PB}))$$
$$- (\overrightarrow{PC} - \overrightarrow{PB} - \overrightarrow{PA} - (\overrightarrow{PC} - \overrightarrow{PB})) \cdot (-\overrightarrow{PB} - (\overrightarrow{PC} - \overrightarrow{PB})) = 0.$$

Simplifying further, we have
$$\overrightarrow{PD} \cdot \overrightarrow{PB} - \overrightarrow{PA} \cdot \overrightarrow{PC} = 0.$$

So, we conclude that points A, B, C, and D are concyclic if and only if points C, Q, R, and S are concyclic.

Example 6. In Figure 4.6, points A, B, C, and D lying on a line such that $AB : BC : CD = 2 : 1 : 3$, and circles with diameters AC and BD are drawn, intersecting at points E and F. Find the value of $\frac{ED}{EA}$.

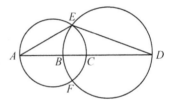

Figure 4.6

Proof. let $E = 0$, $C = \frac{A+D}{2}$, $B = \frac{2A+D}{3}$, and $|ED|^2 = k_0|EA|^2$. We have
$$(D^2 - k_0 A^2) + k_1 A \frac{A+D}{2} + k_2 D \frac{2A+D}{3} = 0,$$
which simplifies to
$$A^2 \left(-k_0 + \frac{k_1}{2}\right) + D^2 \left(1 + \frac{k_2}{3}\right) + AD \left(\frac{k_1}{2} + \frac{2k_2}{3}\right) = 0.$$
Solving the system of equations, $-k_0 + \frac{k_1}{2} = 1 + \frac{k_2}{3} = \frac{k_1}{2} + \frac{2k_2}{3} = 0$, we obtain $k_0 = 2$, $k_1 = 4$, and $k_2 = -3$. Therefore,
$$(D^2 - 2A^2) + 4A \frac{A+D}{2} - 3D \frac{2A+D}{3} = 0,$$
and $|ED| = \sqrt{2}|EA|$.

If we consider a proof problem as a "complete question with conditions and conclusions", then a calculation problem can be seen as "partially incomplete". By keeping some parts "sufficiently complete", we can determine the "incomplete parts". Undoubtedly, calculation problems are harder than proof problems, and using the identity method only requires introducing one more unknown.

Example 7. In Rt$\triangle ABC$, point D is the midpoint of the hypotenuse AB, and point P is the midpoint of segment CD. Find $\frac{|PA|^2+|PB|^2}{|PC|^2}$. (2012 Jiangxi Provincial College Entrance Examination (Science))

Proof. Identity: let $C = 0$, then

$$\left(\frac{A+B}{4} - A\right)^2 + \left(\frac{A+B}{4} - B\right)^2 - 10\left(\frac{A+B}{4}\right)^2 + 2AB = 0.$$

Let $C = 0$, and suppose

$$\left(\frac{A+B}{4} - A\right)^2 + \left(\frac{A+B}{4} - B\right)^2 - x\left(\frac{A+B}{4}\right)^2 + kAB = 0.$$

We expand this equation to get

$$\frac{1}{16}(10-x)A^2 + \frac{1}{16}(10-x)B^2 + \frac{1}{8}(-6+8k-x)AB = 0.$$

Solving the equation $\frac{1}{16}(10-x) = \frac{1}{16}(10-x) = \frac{1}{8}(-6+8k-x) = 0$ gives $x = 10$ and $k = 2$. Hence, we obtain the identity

$$|PA|^2 + |PB|^2 = 10|PC|^2 \iff CA \perp CB.$$

In vector form,

$$\left(\frac{\overrightarrow{CA}+\overrightarrow{CB}}{4} - \overrightarrow{CA}\right)^2 + \left(\frac{\overrightarrow{CA}+\overrightarrow{CB}}{4} - \overrightarrow{CB}\right)^2$$
$$- 10\left(\frac{\overrightarrow{CA}+\overrightarrow{CB}}{4}\right)^2 + 2\overrightarrow{CA}\cdot\overrightarrow{CB} = 0.$$

In this problem, observing the coefficient of A^2 allows us to quickly determine that $x = 10$.

Exercise 4.1

1. In Figure 4.7, let I be the incentre of $\triangle ABC$ and M be the midpoint of AC, and let W be the midpoint of the arc AB on the circumcircle of $\triangle ABC$ not containing point C. It is known that $\angle AIM = 90°$, then find the ratio $CI : IW$. (2017 Saurege Geometry Olympiad 9th Grade Final)

2. In Figure 4.8, D is a point on the extension of AC, and $AB = AC = CD$. E, F, and G are respectively the midpoints of AB, AC, and BD. Prove that $FE \perp FG$.

Identity-Based Method 2: Undetermined Coefficients   105

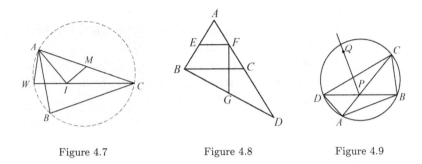

Figure 4.7   Figure 4.8   Figure 4.9

3. In Figure 4.9, let the diagonals AC and BD of cyclic quadrilateral $ABCD$ intersect at point P. Q is the circumcentre of $\triangle CDP$. Prove that $PQ \perp AB$.
4. In Figure 4.10, in $\triangle ABC$, where $AB = AC$, M is the midpoint of BC and N is the midpoint of AM. MD is perpendicular to CN at point D. Prove that $AD \perp BD$.
5. In Figure 4.11, $\triangle ABC$ is a right-angled triangle. A perpendicular line is drawn from point D on BC to AB, with the foot of the perpendicular denoted as E. This line intersects the extension of side AC at point F and the circumcircle of $\triangle ABC$ at point G. Prove that $EG^2 = ED \cdot EF$.

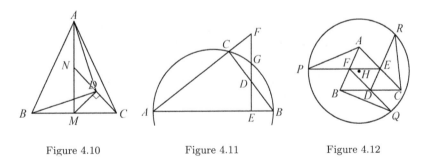

Figure 4.10   Figure 4.11   Figure 4.12

6. In Figure 4.12, H is the orthocentre of $\triangle ABC$. A circle is drawn with centre H such that $\triangle ABC$ is entirely inside the circle. Points D, E, and F are the midpoints of the sides BC, CA, and AB, respectively. Lines EF, FD, and DE are extended to intersect the circle at P, Q, and R, respectively. Prove that $AP = BQ = CR$.

7. In Figure 4.13, AE is a chord of the circumcircle of $\triangle ABC$. D is the midpoint of BC. F is a point on line AE. H is the orthocentre of $\triangle ABC$. Prove that $HF \perp AF \iff DE = DF$.
8. In Figure 4.14, H is the orthocentre of $\triangle ABC$. M is the midpoint of BC. A circle with diameter AH intersects the circumcircle of $\triangle BHC$ at point X (other than H). Prove that points A, X, and M are collinear.

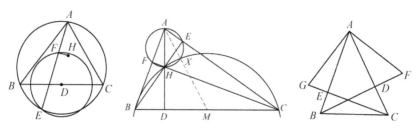

Figure 4.13   Figure 4.14   Figure 4.15

9. In Figure 4.15, BD and CE are altitudes of $\triangle ABC$, and points F and G are taken on the extensions of BD and CE (or on the reverse extensions) such that $BF = AC$ and $CG = AB$. Prove that $AF \perp AG$.
10. We say that two circles are orthogonal if these two circles intersect at two points, and two tangent lines of the circles at each intersection point are perpendicular. Given $\sin^2 A + \sin^2 B + \sin^2 C = 1$, prove that the circumcircle and the nine-point circle of $\triangle ABC$ are orthogonal.
11. In Figure 4.16, BD and CE are altitudes of $\triangle ABC$, and F is the midpoint of BC. Point T is symmetric to point F about line BD. Prove that $ST \perp SH$.

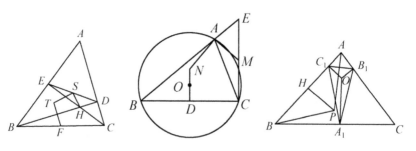

Figure 4.16   Figure 4.17   Figure 4.18

12. In Figure 4.17, O is the circumcentre of $\triangle ABC$. D is the midpoint of BC, and N is the point symmetric to D about O. Point E lies on the extension of side BA. M is the midpoint of CE. Prove that $AM \perp AN \iff CB \perp CE$.

13. In Figure 4.18, consider that O is an interior point of acute $\triangle ABC$. The projections of O onto BC, CA, and AB are A_1, B_1, and C_1, respectively. Perpendiculars are drawn from points A and B to lines B_1C_1 and A_1C_1, respectively, and they intersect at point P. Prove that $CP \perp A_1B_1$.

14. In Figure 4.19, in $\triangle ABC$, where M is the midpoint of BC, construct equilateral $\triangle BAP$ outwardly on side AB and equilateral $\triangle ACK$ outwardly on side AC. Let Q be the centre of $\triangle ACK$. Prove that $MP \perp MQ$.

15. In Figure 4.20, in parallelogram $ABCD$, altitude $BH \perp AD$, and M is on BH such that $MC = MD$. K is the midpoint of AB. Prove that $KM \perp KD$.

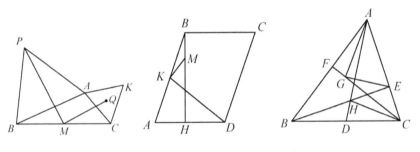

Figure 4.19   Figure 4.20   Figure 4.21

16. In Figure 4.21, BE and CF are altitudes of $\triangle ABC$, D is the midpoint of BC, and point G is on line CF. AD intersects BE at H. Prove that $AD \perp EG \iff AG \perp HC$.

17. In Figure 4.22, in the right trapezium $ABCD$, where $AD \parallel BC$ and $AB \perp BC$, point E is the midpoint of AB, and point F is the midpoint of BC. Furthermore, $AD = DC$. Prove that $CE \perp DF$.

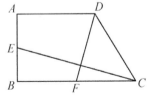

Figure 4.22

18. Given that E and F are the midpoints of sides AD and CD, respectively, of rhombus $ABCD$, and $BE \perp AF$, prove that $ABCD$ is a square. (*Mathematical Bulletin* Problem Solving 1159)
19. In Figure 4.23, in the right triangular prism $ABC - A_1B_1C_1$, where $AB_1 \perp BC_1$, $BC_1 \perp CA_1$, and $CA_1 \perp AB_1$. Prove that the prism is a regular triangular prism. (Problem Solving 1286 in *Mathematical Bulletin*)
20. In Figure 4.24, points C and D lie on the semicircle with diameter AB. A line is drawn from point D perpendicular to the tangent at point C, and the foot of the perpendicular is E. Prove that $AE^2 + BE^2 = \frac{AB}{DE}CD^2 + 2EC^2$. (Problem Solving 1130 in *Mathematical Bulletin*)

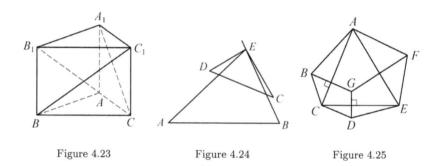

Figure 4.23   Figure 4.24   Figure 4.25

21. In Figure 4.25, in convex hexagon $ABCDEF$, $AB = AF$, $BC = CD$, and $DE = EF$. A line BE is drawn from point B perpendicular to AC, and a line DG is drawn from point D perpendicular to CE. Let G be the point where these two lines intersect inside $\triangle ACE$. Prove that $FG \perp AE$. (Problem Solving 1606 in *Mathematical Bulletin*)
22. In Figure 4.26, if the projections of point P lies on the extensions of the sides BC, CA, and AB of $\triangle ABC$ are X, Y, and Z, respectively. Prove that the perpendiculars drawn from the midpoints of YZ, ZX, and XY to the lines BC, CA, and AB, respectively, intersect at a common point M.
23. In Figure 4.27, O and H are the circumcentre and orthocentre of $\triangle ABC$, respectively. The reflections of O about BC, CA, and AB are denoted as X, Y, and Z, respectively. Prove that lines AX, BY, and CZ intersect at a point P and P is the midpoint of OH. If D, E, and F are the midpoints of BC, CA, and AB, respectively, then P is

the circumcentre of $\triangle DEF$. (Problem Solving 1130 in *Mathematical Bulletin*)

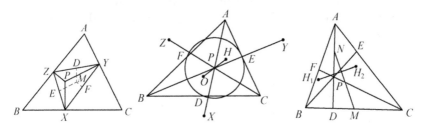

Figure 4.26   Figure 4.27   Figure 4.28

24. In Figure 4.28, in $\triangle ABC$, where AD is the altitude from vertex A on BC and D is on BC, point P lies on AD. Line BP intersects AC at point E, and line CP intersects AB at point F. The orthocentres of $\triangle BCF$ and $\triangle BCE$ are H_1 and H_2, respectively. The midpoints of BC and AP are M and N, respectively. Prove that $H_1H_2 \perp MN$.

25. In Figure 4.29, $\triangle ABC$ is given, and D, E, and F are points on sides BC, CA, and AB, respectively, such that A, B, D, and E are concyclic, and C, A, F, and D are concyclic. O is the circumcentre of $\triangle AFE$. Prove that $OD \perp BC$.

26. In Figure 4.30, AB is the diameter of circle O, and chord $CD \parallel AB$. An extension of AD intersects the tangent line of circle O passing B at point E, and G is the point symmetric to A about C. Prove that $GE \perp GA$.

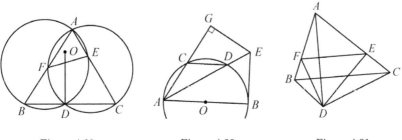

Figure 4.29   Figure 4.30   Figure 4.31

27. In Figure 4.31, point D is located outside of $\triangle ABC$, and $AD \perp BC$. Line $EF \parallel BC$, and $BD \perp DE$. Prove that $DF \perp DC$.

28. In Figure 4.32, CD is an altitude of $\triangle ABC$, and AM is a median. An extension of AM intersects the circumcircle at point E, and a line is drawn from E perpendicular to AE, intersecting the extension of DC at point F. Prove that $FC \cdot CD = AB \cdot DB$. (Provided by Ye Zhonghao)

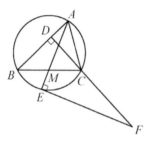

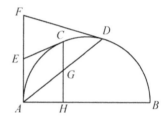

Figure 4.32   Figure 4.33

29. In Figure 4.33, a semicircle is drawn with AB as its diameter, and points C and D lie on the semicircle. Tangent lines CE and DF intersect the tangential line starting at point A at points E and F, respectively. CH is perpendicular to AB, with H as the foot of the altitude. AD intersects CH at G. Prove that $\frac{FE}{EA} = \frac{CG}{GH}$.
30. In $\triangle ABC$, where AD is perpendicular to BC at point D, G is the centroid, and GH is perpendicular to BC at point H. Prove that $BD^2 - CD^2 = 3(BH^2 - CH^2)$.
31. In Figure 4.34, for quadrilateral $ABCD$, where the perpendicular bisectors of AB and CD intersect at point P, and the perpendicular bisectors of AD and BC intersect at point Q, M and N are the midpoints of AC and BD, respectively. Prove that $PQ \perp MN$.

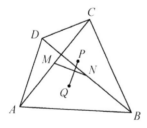

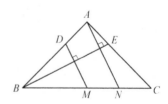

Figure 4.34   Figure 4.35

Identity-Based Method 2: Undetermined Coefficients

32. In Figure 4.35, in the right-angled $\triangle ABC$, where $AB = AC$, $AD = AE$, and $AN \perp BE$, if $DM \perp BE$, prove that $MN = NC$. (32nd IMO China Training Team Training Problem)

33. If $\overrightarrow{OA}, \overrightarrow{OB}, \overrightarrow{OC}$, and $\overrightarrow{OD}$ are unit vectors and $\overrightarrow{OA}+\overrightarrow{OB}+\overrightarrow{OC}+\overrightarrow{OD} = 0$, prove that $|AB| = |CD|$.

34. In Figure 4.36, H is the orthocentre of $\triangle ABC$, and a circle with diameter AC intersects the circumcircle of $\triangle HAB$ at points A and K. Prove that the line CK bisects segment BH. (2017 Pan African Math Olympiad)

35. In Figure 4.37, point H is the orthocentre of $\triangle ABC$ and a circle with diameter AC intersects the circumcircle of $\triangle HAB$ at points A and K. The midpoint of segment HB is X. Prove that points C, K, and X are collinear.

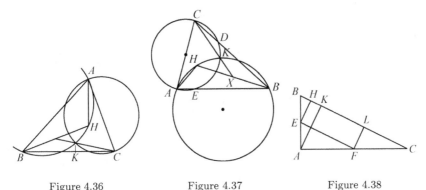

Figure 4.36 Figure 4.37 Figure 4.38

36. In Figure 4.38, AK is the altitude of $\text{Rt}\triangle ABC$ on BC, and the side LH of rectangle $EFLH$ lies on BC. Points E and F lie on sides AB and AC, respectively. Prove that $CK \cdot CH - BK \cdot BL = AC^2 - AB^2$.

37. In Figure 4.39, AD is one of the parallel sides of trapezium $ABCD$. The circumcentre P of $\triangle ABC$ is on BD. Prove that the circumcentre Q of $\triangle ABD$ is on line AC.

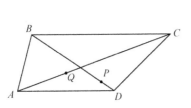

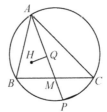

Figure 4.39 Figure 4.40

38. In Figure 4.40, H is the orthocentre of $\triangle ABC$. AM intersects the circumcircle at point P. P and Q are symmetric about M. Prove that $HQ \perp AM$.

39. In Figure 4.41, in $\triangle ABC$, $AD \perp BC$ at D, and E and F are points on sides AC and AB, respectively, such that $CE = CD$ and $BF = BD$. A line parallel to EF is drawn through A and intersects the circumcircle of $\triangle ABC$ at point P. Prove that $PE = PF$.

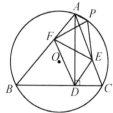

Figure 4.41

4.2 Applications

Example 1. In Figure 4.42, in $\triangle ABC$, H is the orthocentre, and P is a point on the altitude AD such that $PD^2 = AD \cdot HD$. Connect BP and CP. Prove that $PB \perp PC$. (1979 Heilongjiang Mathematics Competition)

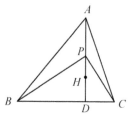

Figure 4.42

Proof 1. Let $P = 0$, then $BC + [(A - D)(H - D) - D^2] + (A - H)(B - D) + (C - H)(A - B) - 2A\left(\frac{B+C}{2} - D\right) = 0$.

Proof 2. Let $D = 0$, then $(P - B)(P - C) - (P^2 - AH) + PB + PC - AB - HC + (B - H)(A - C) = 0$.

Note. Theoretically, any point can be chosen as the origin, but for simplicity, it is best to select a vertex or an intersection point of multiple lines. The choice of the origin can vary, but selecting $P = 0$ is a natural choice when the goal is to prove $PB \perp PC$, and the condition $PD^2 = AD \cdot HD$ hints at considering $D = 0$. The actual choice may depend on the problem structure.

Example 2. In Figure 4.43, O is the circumcentre of $\triangle ABC$, D is the midpoint of AB, E is the centroid of $\triangle ACD$, and $AB = AC$. Prove that $OE \perp CD$. (1983 British Mathematical Olympiad)

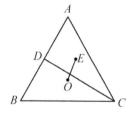

Figure 4.43

Proof. $6\left(O - \frac{A+C+\frac{A+B}{2}}{3}\right)\left(C - \frac{A+B}{2}\right) + 3\left(O - \frac{B+C}{2}\right)(B - C) + 3\left(O - \frac{C+A}{2}\right)(A - C) + [(A - B)^2 - (A - C)^2] = 0.$

From the identity, we can derive a new proposition. Let O be the circumcentre of $\triangle ABC$, D is the midpoint of AB, E is the centroid of $\triangle ACD$, and $OE \perp CD$. Prove that $AB = AC$.

From a constructive perspective, it is natural to have $AB = AC \Rightarrow OE \perp CD$. Conversely, without knowing $AB = AC$, it is challenging to ensure that $OE \perp CD$. From a problem-solving perspective, we are more accustomed to using $AB = AC$ since isosceles triangles have many familiar properties. On the other hand, $OE \perp CD$ might seem less intuitive and challenging to approach. This reminds one of the Steiner–Lehmus theorem, where proving $AB = AC \Rightarrow OE \perp CD$ is more straightforward, but proving that "in a triangle where both internal angle bisectors are equal, it is isosceles" can be more challenging. Hence, in traditional geometry studies, properties such as $AB = AC \Leftarrow OE \perp CD$ are less common. However, using the method of identities makes it easy to uncover such properties.

Example 3. In Figure 4.44, O and H are the circumcentre and orthocentre of $\triangle ABC$, respectively. $AD \perp BC$ at D, and the perpendicular from A to BC extended intersects CB extended at E. Prove that the circumcircle of $\triangle ADE$ passes through the midpoint of OH. (2012 China Western Mathematics Invitation Tournament)

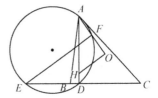

Figure 4.44

Proof. Let $O = 0$, then $H = A + B + C$, $\left(\frac{A+B+C}{2} - A\right)\left(\frac{A+B+C}{2} - E\right) + \frac{1}{4}\left[(E - A)^2 - E^2\right] - \frac{B+C}{2}\left(\frac{B+C}{2} - E\right) = 0.$

Note. The circumcircle of $\triangle ADE$ has AE as its diameter. Also, $\frac{A+B+C}{2}$ lies on the circle that has the diameter AE.

Example 4. In Figure 4.45, H is the orthocentre of $\triangle ABC$. Draw a circle with BC as the diameter. AP and AQ are tangents, and P and Q are the points of tangency. Prove that P, Q, and H are collinear. (1996 China Mathematical Olympiad)

Proof. Let $H = 0$, then

$$2P\left(A - \frac{B+C}{2}\right) + 2(P-A)\left(P - \frac{B+C}{2}\right)$$
$$- 2(P-B)(P-C) - B(A-C) - C(A-B) = 0.$$

Note. This proves $AO \perp HP$; similarly, $AO \perp HQ$.

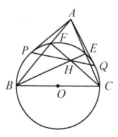
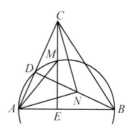

Figure 4.45   Figure 4.46

Example 5. In Figure 4.46, in acute $\triangle ABC$, BD and CE are altitudes on AC and AB, respectively. A circle with diameter AB intersects CE at M. Point N is taken on BD such that $AN = AM$. Prove that $AN \perp CN$. (The 3rd Northern Mathematical Olympiad Invitation Tournament)

Proof. $(N-C)(N-A) + [(A-M)^2 - (A-N)^2] - (A-C)(N-B) - (M-A)(M-B) - (C-M)(A-B) = 0.$

Example 6. In Figure 4.47, in acute $\triangle ABC$, where M is the circumcentre, a circle passing through points A, B, and M intersects line BC at point P and line AC at point Q. Prove that $CM \perp PQ$. (The 34th IMO Turkey National Final)

Proof. Let $M = 0$, then

$$2C(P-Q) + [(C-B)(C-P) - (C-A)(C-Q)] - 2\frac{A+C}{2}(C-Q)$$
$$+ 2\frac{B+C}{2}(C-P) = 0.$$

Note. The conclusion obviously holds even when M is not on the circle.

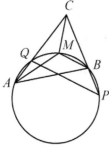 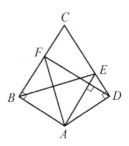

Figure 4.47   Figure 4.48

Example 7. In Figure 4.48, in quadrilateral $ABCD$, E and F are points on DC and CB, respectively, and $AB = AD$. $DF \perp AE$, $AB \perp BC$, and $AD \perp DC$. Prove that $AF \perp BE$. (1995 Russian Federation Regional Competition)

Proof. Let $A = 0$, then

$$F(B - E) + E(F - D) - B(F - B) + D(E - D) + (D^2 - B^2) = 0.$$

Example 8. In Figure 4.49, in trapezium $ABCD$ with $AB \parallel CD$, AC intersects BD at E. F and G are the orthocentres of $\triangle EBC$ and $\triangle EDA$, respectively, and H is the midpoint of FG. Prove that $EH \perp AB$. (1997 Austria-Poland Mathematical Competition)

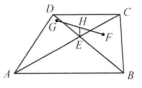

Figure 4.49

Proof. Let $E = 0$, then $2\frac{G+F}{2}(A - B) + A(D - G) + A(B - F) - B(C - F) - B(A - G) - (AD - BC) = 0$.

Example 9. In Figure 4.50, in acute $\triangle ABC$, H is the orthocentre, which is the common intersection of the altitudes AA_1, BB_1, and CC_1. M and N are the midpoints of segments BC and AH, respectively. Prove that MN is the perpendicular bisector of segment B_1C_1. (2013 Belarus Math Olympiad)

Proof.
$$\left(\frac{B+C}{2} - C_1\right)^2 - \left(\frac{B+C}{2} - B_1\right)^2$$
$$+ (B_1 - B)(B_1 - C) - (C_1 - C)(C_1 - B) = 0,$$
$$\left(\frac{A+H}{2} - C_1\right)^2 - \left(\frac{A+H}{2} - B_1\right)^2$$
$$+ (B_1 - H)(B_1 - A) - (C_1 - H)(C_1 - A) = 0.$$

Note. These two identities imply that M and N, respectively, lie on the perpendicular bisector of segment B_1C_1.

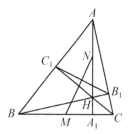
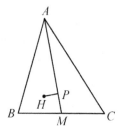

Figure 4.50   Figure 4.51

Example 10. In Figure 4.51, let H be the orthocentre of acute $\triangle ABC$ and M be the midpoint of BC. Draw the perpendicular from H to AM, and let P be the foot of the perpendicular. Prove that $AM \cdot PM = BM^2$. (2011 Japan Mathematical Olympiad)

Proof.
$$\left[\left(A - \frac{B+C}{2}\right)\left(P - \frac{B+C}{2}\right) - \left(B - \frac{B+C}{2}\right)^2\right]$$
$$+ (H - P)\left(A - \frac{B+C}{2}\right) - \frac{1}{2}(H - C)(A - B) - \frac{1}{2}(H - B)(A - C) = 0.$$

Note. Letting $\frac{B+C}{2} = 0, C = -B$ would simplify the proof.

Example 11. In Figure 4.52, there is a rectangle $ABCD$ with point E on the circumcircle. The tangent to the circle at point A intersects line DE at F. Lines BE and AC intersect at K. Prove that $FK \perp BC$. (The 14th Saurege Geometry Olympiad 8th Grade)

Proof. Let $K = 0$, then $F(B - C) + (AC - BE) - (F - E)B + (F - A)C = 0$.

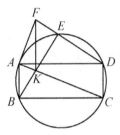

Figure 4.52

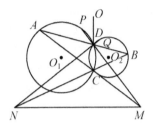

Figure 4.53

Example 12. In Figure 4.53, circles $\odot O_1$ and $\odot O_2$ intersect at C and D. A line passing through D intersects $\odot O_1$ and $\odot O_2$ at A and B, respectively. Point P lies on arc AD of $\odot O_1$. Line PD intersects the extension of segment AC at M, and point Q lies on arc BD of $\odot O_2$. Line QD intersects the extension of segment BC at N. O is the circumcentre of $\triangle ABC$. Prove that $OD \perp MN$ if and only if points P, Q, M, and N are concyclic. (2007 7th China Western Mathematical Olympiad)

Proof.

$$2(O - D)(M - N) + [(D - Q)(D - N) - (D - P)(D - M)]$$
$$- [(O - N)^2 - (O - A)^2 - (N - C)(N - B)]$$
$$+ [(O - M)^2 - (O - A)^2 - (M - C)(M - A)]$$
$$+ [(M - C)(M - A) - (M - D)(M - P)]$$
$$+ [(N - D)(N - Q) - (N - C)(N - B)] = 0.$$

Example 13. In Figure 4.54, in $\triangle ABC$, where $\angle A$ is a right angle, point D lies on BC such that $BD = 4DC$. A circle passes through point C and intersects AC at F and is tangent to AB at the midpoint G. Prove that $AD \perp BF$. (1999 China National Junior Secondary School Mathematics Competition)

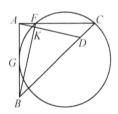

Figure 4.54

Proof. Let $A = 0$, then
$$5\left(-\frac{B+4C}{5}\right)(B-F) + 4BC + 4\left[\left(\frac{B}{2}\right)^2 - CF\right] - BF = 0.$$

Example 14. In Figure 4.55, in $\triangle ABC$, where H is the orthocentre and M is the midpoint of CA, a line BL is drawn, which is tangent to the circumcircle of $\triangle ABC$ at B, and $HL \perp BL$. Prove that $\triangle MBL$ is isosceles. (1999 National Junior High School Mathematics Competition)

Proof. Let the circumcentre of $\triangle ABC$ be the origin, and $H = A+B+C$:
$$\left(\frac{A+C}{2} - B\right)^2 - \left(\frac{A+C}{2} - L\right)^2$$
$$+ (A+B+C-L)(B-L) - 2(B-L)B = 0.$$

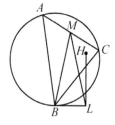
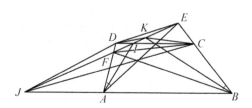

Figure 4.55　　　　　　　　　Figure 4.56

Example 15. In Figure 4.56, in trapezium $ABCD$ with $AB \parallel DC$, point E lies on the extension of BC, and point F lies on AD. Additionally, $\angle EAD = \angle CBF$. Line EF intersects CD and BA at I and J, respectively, and K is the midpoint of EF (distinct from I and J). Prove that points $K, I, A,$ and B are concyclic if and only if points $K, D, J,$ and C are concyclic. (2009 Benelux Mathematical Olympiad)

Proof.
$$\left[(J-A)(J-B) - (J-I)\left(J - \frac{E+F}{2}\right)\right]$$
$$+ \left[(I-J)\left(I - \frac{E+F}{2}\right) - (I-C)(I-D)\right]$$
$$+ [(J-E)(J-F) - (J-A)(J-B)]$$
$$+ [(I-C)(I-D) - (I-E)(I-F)] = 0.$$

Note. The key is to note that ∠DFE = ∠ABC = ∠DCE, which makes D, F, C, and E concyclic.

Example 16. In Figure 4.57, in △ABC, where O is the circumcentre and H is the orthocentre, a line through O parallel to BC intersects AB at K. M is the midpoint of AH. Prove that ∠CMK = 90°. (2018 Hong Kong IMO Team Selection Test-Round 4)

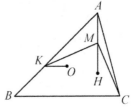

Figure 4.57

Proof. Let $O = 0$: $\left(\frac{A+B+C+A}{2} - K\right)\left(\frac{A+B+C+A}{2} - C\right) + \frac{1}{4}(C^2 - B^2) - 2\frac{A+B}{2}(A-K) - K\frac{B+C}{2} = 0.$

Example 17. In Figure 4.58, in quadrilateral $ABCD$, inscribed in the circle with diameter AC, points E and F lie on segments CD and BC, respectively, such that $AE \perp DF$ and $AF \perp BE$. Prove that $AB = AD$.

Proof. Let $A = 0$:

$$(B^2 - D^2) - D(E - D) - B(B - F) - E(F - D) - F(B - E) = 0.$$

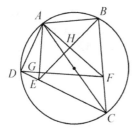

Figure 4.58

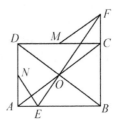

Figure 4.59

Example 18. In Figure 4.59, let $ABCD$ be a rectangle with centre O, and $AB \neq BC$. The perpendicular from O to BD cuts the lines AB and BC at E and F, respectively. Let M and N be the midpoints of CD and AD, respectively. Prove that $FM \perp EN$. (2008 Romania National Olympiad)

Proof.

$$\left(F - \frac{A+C-B+C}{2}\right)\left(E - \frac{A+C-B+A}{2}\right)$$
$$- \left(\frac{A+B}{2} - E\right)\left(\frac{B+C}{2} - F\right)$$
$$+ 2\left(B - \frac{A+C}{2}\right)\left(\frac{A+C}{2} - \frac{E+F}{2}\right) = 0.$$

Example 19. In Figure 4.60, in acute $\triangle ABC$, AH is an altitude, and AM is a median. Points X and Y lie on lines AB and AC, respectively, such that $AX = XC$ and $AY = YB$. N is the midpoint of XY. Prove that $NM = NH$. (2018 Ukraine Mathematical Olympiad)

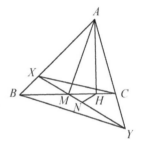

Figure 4.60

Proof. $\left(\frac{X+Y}{2} - \frac{\frac{B+C}{2}+H}{2}\right)(B - C) + \frac{1}{2}(B - C)(H - A) + \left(X - \frac{A+C}{2}\right)\left(\frac{A+C}{2} - Y\right) - \left(Y - \frac{A+B}{2}\right)\left(\frac{A+B}{2} - X\right) = 0.$

From the identity, a more general proposition can be derived. In Figure 4.61, in $\triangle ABC$, where $AH \perp BC$, AM is the median, and points X and Y lie on lines AB and AC, respectively, such that $AX = XC$ and $AY = YB$. N is the midpoint of XY, and K is the midpoint of MH. Prove that $NK \perp BC$.

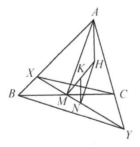

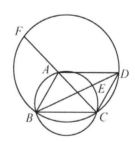

Figure 4.61   Figure 4.62

Identity-Based Method 2: Undetermined Coefficients 121

Example 20. In Figure 4.62, in quadrilateral $ABCD$, a circle passing through points A, B, and C intersects BD at E, and a circle passing through points B, C, and D intersects the extension of CA at F. Prove that $BD \cdot BE = AC \cdot CF$.

Proof. Let $\frac{A+C}{2} = 0$, $D = -B$, and $A = -C$:
$$[(B-D)(B-E) - (C-A)(C-F)] - 2(AC - BE) - 2(FC - BD) = 0.$$

Example 21. Consider five points A, B, C, D, and E distributed in order on circle O, where $AB \parallel ED$. Prove that $AC^2 = BD^2 + CE^2 \iff \angle ABC = 90°$. (2005 British Mathematical Olympiad)

Proof. Let $O = 0$, then
$$[(A-C)^2 - (B-D)^2 - (C-E)^2] + 2(E-A)(E-C)$$
$$+ (B^2 - A^2 + D^2 - E^2) - 2\frac{A+B}{2}(D-E) + 2\frac{D+E}{2}(A-B) = 0.$$

Note. This implies $\angle ABC = 90°$ if and only if AC is the diameter, which is equivalent to $\angle AEC = 90°$.

Example 22. In Figure 4.63, in parallelogram $ABCD$, a line through A intersects CB at E and CD at F. Prove that $CB \cdot CE + CD \cdot CF = AC^2 + AE \cdot AF$.

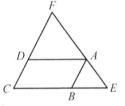

Figure 4.63

Proof. Let $C = 0$, then $A = B + D$, and
$$[BE + DF - (B+D)^2 + (B+D-E)(B+D-F)]$$
$$+ [(F-D)(F-E) - (F-(B+D))F] = 0.$$

Note. $(F-D)(F-E) - (F-(B+D))F = 0$ involves the use of triangle similarity, i.e.,
$$\frac{|FD|}{|FC|} = \frac{|FA|}{|FE|}.$$

Example 23. In Figure 4.64, in pentagon $ABCDE$, $\angle ABC = \angle AED = 90°$, $\angle BAC = \angle EAD$, and F is the midpoint of CD. Prove that $BF = EF$.

Proof. Let $A = 0$, then
$$\left(B - \frac{C+D}{2}\right)^2 - \left(E - \frac{C+D}{2}\right)^2$$
$$- B(B-C) + E(E-D) + (BD - CE) = 0.$$

122 Solving Problems in Point Geometry

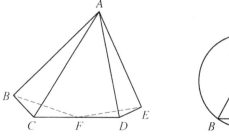
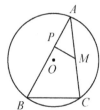

Figure 4.64   Figure 4.65

Example 24. In Figure 4.65, it is known that BC is a chord on the circle, and point A moves along the circle. M is the midpoint of AC, and a line through M perpendicular to AB intersects AB at P. Find the locus of point P.

Proof. Let the centre O of the circle be the origin. Then,

$$\left(P - \frac{-C+B}{2}\right)(P-C) - \frac{B+C}{2}(B-C) - \frac{A+B}{2}(P-B)$$
$$+ (P-B)\left(\frac{A+C}{2} - P\right) = 0.$$

The equation $\left(P - \frac{-C+B}{2}\right)(P-C) = 0$ implies that the locus of point P is a circle. Note that the geometric meaning of $-C$ is the antipodal point of C about O.

Example 25. In Figure 4.66, chords AC and BD intersect at point P, and perpendiculars are drawn at points C and D on AC and BD, respectively, intersecting at point Q. Prove that $AB \perp PQ$. (The 1st Saurege Geometry Olympiad)

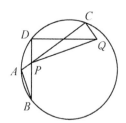

Proof. Let $P = 0$:

$$Q(A-B) - (AC - BD) + A(C-Q) - B(D-Q) = 0.$$

Figure 4.66

Example 26. In Figure 4.67, in parallelogram $ABCD$, diagonals AC and BD intersect at point O. Draw a tangent to the circumcircle of $\triangle BOC$

passing through O, and let it intersect CB at point F, and line BC intersects the circumcircle of $\triangle FOD$ at a point G other than F. Prove that $AG = AB$.

Proof.
$$2\left(A - \frac{G+B}{2}\right)(F-B) + 2\left[\left(F - \frac{A+C}{2}\right)^2 - (F-B)(F-C)\right]$$
$$+ \left[(B-G)(B-F) - \left(B - \frac{A+C}{2}\right)(B-(A+C-B))\right] = 0.$$

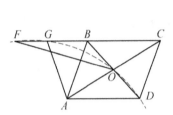

Figure 4.67 Figure 4.68

Example 27. In Figure 4.68, in $\triangle ABC$, O is the circumcentre, and D, E, and F are on BC, CA, and AB, respectively. $DE \perp CO$, and $DF \perp BO$. Let K be the circumcentre of $\triangle AFE$. Prove that $DK \perp BC$. (2012 European Girls' Mathematical Olympiad)

Proof. Let $O = 0$:
$$(K-D)(B-C) - (F-D)B + (E-D)C - \frac{B+A}{2}(A-F)$$
$$- \frac{C+A}{2}(E-A) + \left(K - \frac{F+A}{2}\right)(A-B)$$
$$- \left(K - \frac{E+A}{2}\right)(A-C) = 0,$$
$$(B-D)(B-C) - (B-F)(B-A) - 2(F-D)B$$
$$+ 2\frac{B+C}{2}(B-D) - 2\frac{B+A}{2}(B-F) = 0.$$

It is also found that points D, C, F, and A are concyclic. By symmetry, points B, D, E, and A are concyclic.

Example 28. As shown in Figure 4.69, let H be the orthocentre of $\triangle ABC$. Points D, E, and F are the midpoints of sides BC, CA, and AB, respectively. Points L, M, and N are on lines BC, CA, and AB, respectively, such that AL, BM, and CN are perpendicular to DH, EH, and FH, respectively. Prove that the points L, M, and N are collinear, and the line is perpendicular to the Euler line.

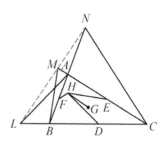

Figure 4.69

Proof. Let $O = 0$, then $H = A + B + C$, and

$$(L - M)(A + B + C) - (L - A)\left(A + B + C - \frac{B+C}{2}\right) + \frac{B+C}{2}(C - L)$$
$$+ (M - B)\left(A + B + C - \frac{A+C}{2}\right) - \frac{A+C}{2}(C - M) - (A^2 - B^2) = 0.$$

Therefore, $LM \perp OH$. Similarly, $LN \perp OH$. So, points L, M, and N are collinear, and this line is perpendicular to the Euler line.

Note. In this method, we use $L = uB + (1-u)C$ for precise positioning. However, this does not restrict L to lie exactly on BC. The equations $(L - A)\left(A + B + C - \frac{B+C}{2}\right) = 0$ and $\frac{B+C}{2}(C - L) = 0$ do not fully restrict L to lie on BC. This shows that the conclusion holds without the restriction of L lying on BC. In some problems, it is necessary to assume that L lies on BC, and in such cases, we have to use $L = uB + (1-u)C$.

Example 29. In Figure 4.70, in acute $\triangle ABC$, a circle with diameter AH is drawn. This circle intersects sides AB and AC at points M and N (other than A), respectively. Perpendiculars to MN are drawn from A, B, and C, resulting in lines l_A, l_B, and l_C, respectively. Prove that lines l_A, l_B, and l_C are concurrent.

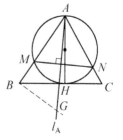

Figure 4.70

Proof. Let $A = 0$. The circumcentre of $\triangle ABC$ is O. Then, $2O(M - N) - M(2O - B) + N(2O - C) - B(M - H) + C(N - H) - H(B - C) = 0$. The identity shows that $AO \perp MN$, and O lies on l_A. Similarly, O lies on l_B and l_C, so lines l_A, l_B, and l_C are concurrent.

Example 30. In Figure 4.71, quadrilateral $ABCD$ is inscribed in circle O, diagonals AC and BD intersect at point P. Let the circumcentres of $\triangle ABP$, $\triangle BCP$, $\triangle CDP$, and $\triangle DAP$ be O_1, O_2, O_3, and O_4, respectively. Prove that lines OP, O_1O_3, and O_2O_4 are concurrent. (1990 China National High School Mathematics League Second Trial)

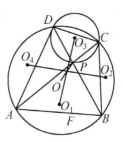

Figure 4.71

Proof. Let $O = 0$. Then, $2\left(P - O_1 - \frac{P+D}{2}\right)(D - B) - 2\left(O_1 - \frac{P+B}{2}\right)(B - D) - (B^2 - D^2) = 0$. This shows that the vertices of parallelogram OO_1PK lie on the perpendicular bisector of PD. Due to symmetry, K also lies on the perpendicular bisector of PC. Therefore, K is the circumcentre of $\triangle CDP$, and $K = O_3$. The midpoint of OP is on O_1O_3. Similarly, the midpoint of OP is on O_2O_4. Thus, OP, O_1O_3, and O_2O_4 are concurrent.

Example 31. In Figure 4.72, point P lies on hypotenuse BC of $\text{Rt}\triangle ABC$, and Q is the midpoint of PC. A line is drawn through P perpendicular to BC, intersecting AB at R. The midpoint of AR is H. A ray HN on the same side as C is drawn perpendicular to AB. Prove that there exists a point G on ray HN such that $AG = CQ$ and $BG = BQ$. (2002 China National High School Mathematics League)

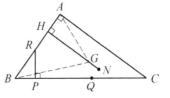

Figure 4.72

Proof. Since $BC > AB$ and $BR > BP$, we have $AH = \frac{AB-BR}{2} < \frac{BC-BP}{2} = CQ$. Therefore, a circular arc with centre A and radius CQ intersects HN at a point G, so $AG = CQ$.

Let $A = 0$. Then,
$$\left[\left(B - \frac{C+P}{2}\right)^2 - (B-G)^2\right] + \left[G^2 - \left(C - \frac{C+P}{2}\right)^2\right]$$
$$+ (B-R)C + (P-R)(B-C) - 2B\left(G - \frac{R}{2}\right) = 0.$$

The explanation from the identity can only elucidate
$$\left[\left(B - \frac{C+P}{2}\right)^2 - (B-G)^2\right] + \left[G^2 - \left(C - \frac{C+P}{2}\right)^2\right] = 0.$$

So, it is still necessary to supplement the explanation with $AG = CQ$.

Example 32. In Figure 4.73, O is the circumcentre of $\triangle ABC$, and H is the intersection point of three altitudes with feet D, E, and F. Lines ED and AB intersect at M, and lines FD and AC intersect at N. Prove that $OB \perp DF$, $OC \perp DE$, and $OH \perp MN$. (2001 National Secondary School Mathematics League)

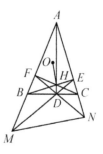

Figure 4.73

Proof 1. Let $O = 0$, then $2B(D-F)+(B+C)(B-D)+(A+B)(F-B)+(A-D)(B-C)+(C-F)(A-B) = 0$.

Proof 2. Let $O = 0$, then $2B(D-F) + (B+C)(C-D) + (F-B)(A+B+C-F) + (A-F)(C-F) - (C-B)(A+B+C-D) = 0$, Therefore, $OB \perp DF$; similarly, $OC \perp DE$. Let $O = 0$, then $(A+B+C)(M-N) + (B+C)(B-D) + (A-B)(A+B+C-F) + B(D-F) + C(D-M) + (A+B)(F-M) - (A+C)(A-N) - B(F-N) = 0$, so $OH \perp MN$.

Note. The second method uses the properties of the Euler line. The first method is simpler and relies on basic geometric relationships.

Example 33. In Figure 4.74, quadrilateral $APBQ$ is inscribed in circle O with $\angle P = \angle Q = 90°$ and $AP = AQ$. Let X be a variable point on segment PQ. Line AX meet O at point S (other than A). Point T lies on O such that $XT \perp AS$, with T on circle O. Prove that as point X moves along PQ, the midpoint M of segment TS traces a circular path. (2005 United States Mathematical Olympiad)

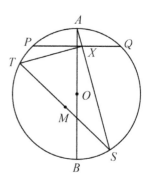

Figure 4.74

Proof. Let $O = 0$, then $4\left(\frac{T+S}{2} - \frac{A}{2}\right)^2 - \left(P - \frac{A}{2}\right)^2 + (3P^2 - Q^2 - T^2 - S^2) - 4A(P - X) + 4\frac{P+Q}{2}\left(\frac{P+Q}{2} - X\right) - 2(A-S)(X-T) + 2[(X-S)(X-A) - (X-P)(X-Q)] = 1$.

Note. This identity shows that the locus of point M is a circular arc with the midpoint of OA as the centre and a radius equal to the distance from the midpoint of OA to P.

Example 34. In Figure 4.75, in acute triangle $\triangle ABC$ with $AB > AC$, points O and H are the circumcentre and orthocentre, respectively. Point M is the midpoint of side BC. Let AM's extension intersect the circumcircle of $\triangle BHC$ at K. Line HK intersects BC at N. Prove that if $\angle BAM = \angle HAN$, then $AN \perp OH$. (2019 China Western Mathematics Invitation Tournament)

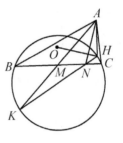

Figure 4.75

Proof. Let $O = 0$, and $H = A + B + C$. Clearly, the circumcentre of $\triangle BHC$ is $B+C$, and its radius squared is A^2. Also, the point symmetric to H about the circumcentre lies on the circumcircle, which means $2(B + C) - (A + B + C) = B + C - A$. Since A, B, C, and $B + C - A$ form a parallelogram, point K lies on AM as well, so $K = B + C - A$.

Now, $\angle AKH = \angle AKC - \angle HKC = \angle KAB - \angle HBC = \angle CAN - \angle HAC = \angle NAH$. Thus, $\triangle AHK$ and $\triangle NHA$ are similar, and $AH^2 = HN \cdot HK$. Using the identity $2(A+B+C)(A-N) - 4\frac{B+C}{2}\left(\frac{B+C}{2} - N\right) + [(A+B+C-A)^2 - (A+B+C-N)(A+B+C-K)] = 0$, we see that $AN \perp OH$.

Generalization. In acute $\triangle ABC$ with $AB > AC$, let O and H be the circumcentre and orthocentre, respectively. Point M is the midpoint of side BC. If the extension of AM intersects the circumcircle of $\triangle BHC$ at point K and line HK intersects BC at point N, prove that

$$AH^2 = HN \cdot HK \iff OH \perp AN.$$

Example 35. Let H be the orthocentre of $\triangle ABC$ and M be the midpoint of side BC. The extension of median AM intersects the circumcircle of $\triangle BHC$ at D. Prove that M is the midpoint of segment AD.

Proof. Let $O = 0$ and $H = A+B+C$. The circumcentre of $\triangle ABC$ is the origin. The circumcentre of $\triangle BHC$ is $B+C$, and its radius is $|OA|$.

Since $M = \frac{B+C}{2}$, and $D_1 = 2M - A = B+C - A$, we see that D_1 also lies on the circumcircle. Thus, $D_1 = 0$, M is the midpoint of segment AD.

Example 36. In Figure 4.76, $\triangle ABC$ is inscribed in circle O, let the internal and external angle bisectors of $\angle A$ intersect BC at M and N, respectively. T is the midpoint of MN. Prove that AT is tangent to circle O.

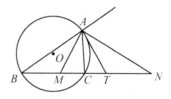

Figure 4.76

Proof 1. Let $A = 0$, then

$$4\frac{M+N}{2}O - 2MN + 4\left(O - \frac{B+C}{2}\right)\left(C - \frac{M+N}{2}\right)$$
$$- 2[O^2 - (O-C)^2] + [(M-B)(N-C) + (N-B)(M-C)] = 0.$$

Proof 2. Let $A = 0$, then

$$\frac{M+N}{2}O = \frac{\frac{bB+cC}{b+c} + \frac{-bB+cC}{-b+c}}{2}O = \frac{b^2BO - c^2CO}{(b-c)(b+c)} = \frac{b^2\frac{c^2}{2} - c^2\frac{b^2}{2}}{(b-c)(b+c)} = 0.$$

Note. The second method is quicker for this problem. This is because in the second method, the points M and N are precisely represented, making the calculations faster. On the other hand, the first method relies on various constraint conditions that need to be eliminated to generate the identity. Although the identity method has various advantages and is incredibly cool, it may not always be the quickest. When solving problems, choose to use it selectively. Remember!

Example 37. In Figure 4.77, in acute $\triangle ABC$ with $AB > AC$, M is the midpoint of BC. The external angle bisector of $\angle BAC$ intersects BC at P. Points K and F lie on line PA such that MF is perpendicular to BC and MK is perpendicular to PA. Prove that $BC^2 = 4PF \cdot AK$. (2017 China Girls Mathematical Olympiad)

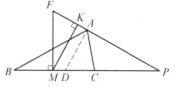

Figure 4.77

Proof 1. Consider angle bisector AD. Clearly, $AD \perp AP$.
Let $B = 0$, $D = \frac{bB+cC}{b+c}$, and $P = \frac{-bB+cC}{-b+c}$:

$$4(P-F)(A-K) = 4(P-F)(D-M) = 4(P-M)(D-M)$$
$$= 4\left(\frac{-bB+cC}{-b+c} - \frac{B+C}{2}\right)\left(\frac{bB+cC}{b+c} - \frac{B+C}{2}\right)$$
$$= C^2.$$

Identity-Based Method 2: Undetermined Coefficients 129

Proof 2. Consider angle bisector AD. Clearly, AD is perpendicular to AP:

$$4(P - F)(A - K) = 4(P - F)(D - M) = 4(P - M)(D - M)$$
$$= \left[4\left(P - \frac{B+C}{2}\right)\left(D - \frac{B+C}{2}\right) - (B - C)^2 \right]$$
$$- 2[(D - B)(P - C) + (P - B)(D - C)] = 0.$$

Note. In this problem, the introduction of point D cleverly eliminates the need for F and K. Calculating F and K directly based on the problem statement would involve a considerable amount of computation. In the second method, the use of identities eliminates the need for the specific coordinates of P and D. When the property of angle bisectors is employed,

$$\frac{|PB|}{|PC|} = \frac{|DB|}{|DC|}.$$

Example 38. In Figure 4.78, given that $\triangle DEF$ is the orthic triangle of $\triangle ABC$, and Z_1, Y_1, and X_1 are the altitudes' feet of $\triangle CDE$, $\triangle BDF$, and $\triangle AEF$, respectively, prove that the perpendiculars drawn from X_1, Y_1, and Z_1 to EF, DF, and DE, respectively, intersect at a common point, K_1.

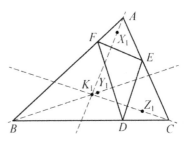

Figure 4.78

Proof.

$$(X_1 - K_1)(E - F) + (Y_1 - K_1)(F - D) + (Z_1 - K_1)(D - E)$$
$$- (A - D)(B - C) - (B - E)(C - A) - (C - F)(A - B)$$
$$- (F - X_1)(A - E) + (E - X_1)(A - F) + (F - Y_1)(B - D)$$
$$- (D - Y_1)(B - F) - (E - Z_1)(C - D) + (D - Z_1)(C - E) = 0.$$

The identity shows that if BY_1 and CZ_1 intersect at point K_1, then K_1 is also on AX_1.

Example 39. In Figure 4.79, consider that $\triangle DEF$ is the orthic triangle of $\triangle ABC$. Z_2, Y_2, and X_2 are the circumcentres of $\triangle CDE$, $\triangle BDF$, and $\triangle AEF$, respectively. Prove that the perpendiculars drawn from X_2, Y_2, and Z_2 to EF, DF, and DE, respectively, intersect at a point, K_2.

Proof.

$$2(X_2 - K_2)(E - F) + 2(Y_2 - K_2)(F - D) + 2(Z_2 - K_2)(D - E)$$
$$+ (X_2 - E)^2 - (X_2 - F)^2 + (Y_2 - F)^2 - (Y_2 - D)^2$$
$$+ (Z_2 - D)^2 - (Z_2 - E)^2 = 0.$$

Note. This shows that K_2 is the circumcentre of $\triangle DEF$.

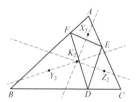 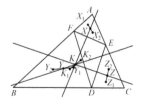

Figure 4.79 Figure 4.80

Example 40. In Figure 4.80, given that $\triangle DEF$ is the pedal triangle of $\triangle ABC$. $Z, Y,$ and X are the midpoints of the altitudes in $\triangle CDE$, $\triangle BDF$, and $\triangle AEF$ (i.e., the nine-point circle centres), respectively. Prove that the perpendiculars drawn from $X, Y,$ and Z to $EF, DF,$ and DE, respectively, intersect at a common point. (provided by Pan Chenghui)

Proof.

$$4\left(\frac{X_1 + X_2}{2} - K\right)(E - F) + 4\left(\frac{Y_1 + Y_2}{2} - K\right)(F - D)$$
$$+ 4\left(\frac{Z_1 + Z_2}{2} - K\right)(D - E)$$
$$- 2(A - D)(B - C) - 2(B - E)(C - A) - 2(C - F)(A - B)$$
$$- 2(F - X_1)(A - E) + 2(E - X_1)(A - F) + 2(F - Y_1)(B - D)$$
$$- 2(D - Y_1)(B - F) - 2(E - Z_1)(C - D)$$
$$+ 2(D - Z_1)(C - E) + (X_2 - E)^2 - (X_2 - F)^2$$
$$+ (Y_2 - F)^2 - (Y_2 - D)^2 + (Z_2 - D)^2 - (Z_2 - E)^2 = 0.$$

Note. Combining the solutions to the two previous problems in a linear manner yields the proof for this problem. It is evident that it is not necessary for the midpoint to be the circumcentre or orthocentre; similar properties hold for points in any proportion.

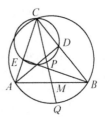

Figure 4.81

Example 41. In Figure 4.81, in acute $\triangle ABC$, let A and B be the feet of the altitudes from A and B to the opposite sides D and E, respectively. Let M be the midpoint of side AB. The line CM intersects the circumcircle of $\triangle CDE$ at the other point P and intersects the circumcircle of $\triangle ABC$ at the other point Q. Prove that $MP = MQ$. (2014 Estonian National Team Selection Exam)

Proof.
$$\left[\left(\frac{A+B}{2} - A\right)\left(\frac{A+B}{2} - B\right) - \left(\frac{A+B}{2} - C\right)\left(\frac{A+B}{2} - (A+B-P)\right)\right]$$
$$- (B-C)(A-D) + \left[(C-P)\left(C - \frac{A+B}{2}\right) - (C-D)(C-B)\right] = 0.$$

Note. Since $\angle CPD = \angle CED = \angle ABC$, it follows that the points M, B, D, and P are concyclic.

Example 42. In Figure 4.82, B is the midpoint of AC, $\angle DBC = \angle DEB$, $PA = PB$, and $AP \perp AD$. Prove that $PE \perp DE$.

Proof.
$$(E-P)(C-D) + \left[(P-A)^2 - \left(P - \frac{A+C}{2}\right)^2\right]$$
$$- (A-P)(A-D) + \left[\left(D - \frac{A+C}{2}\right)^2 - (D-E)(D-C)\right] = 0.$$

Note. This problem involves two transformations. Using angle equality directly in point geometry is inconvenient, so we use the similarity of $\triangle DBC$ and $\triangle DEB$ to transform the problem into $DB^2 = DC \cdot DE$. If we directly calculate $EP \cdot ED = 0$, it is challenging because dealing with E^2 in the equation can be tricky. Additionally, we do not need the condition that

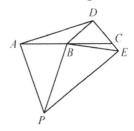

Figure 4.82

points D, C, and E are collinear, so we transform the problem into proving $EP \cdot (C - D) = 0$.

Example 43. In Figure 4.83, in $\triangle ABC$ with $\angle B = 90°$ and $AB > BC$, a semicircle with diameter AB and point C are on the same side of AB. Point P lies on the semicircle such that $BP = BC$, and point Q lies on segment BA such that $AP = AQ$. Prove that the midpoint M of CQ lies on the semicircle. (2008 Dutch National Team Selection Exam)

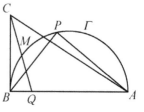

Figure 4.83

Proof. Let $B = 0$:

$$4\left(\frac{C+Q}{2} - A\right)\left(\frac{C+Q}{2} - B\right) - 2P(P - A) + 2C(A - Q)$$
$$+ (P - A)^2 - (Q - A)^2 + (P^2 - C^2) = 0.$$

Example 44. In Figure 4.84, let E be a point on the extension of side AD of rectangle $ABCD$. Line EC intersects the circumcircle of $\triangle ABE$ at the other point F. Line AF intersects line CD at point P. The line passing through point E and parallel to AF intersects the line passing through point C and perpendicular to AF at point Q. Prove that the line PQ is tangent to the circumcircle of $\triangle ABE$.
(2017 Tuymaada Mathematical Olympiad in Russia)

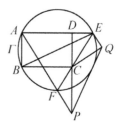

Figure 4.84

Analysis. It is evident that $K = B + E - A$ lies on the circle. We prove that K is the point of tangency between PQ and the circle. We do this in two steps: first, we prove that $KQ \perp KA$, and then we prove that $PQ \perp KA$.

Proof.

$$((B + E - A) - A)((B + E - A) - Q)$$
$$- 2(A - B)(A - E) - (Q - C)(A - F)$$
$$- (B - C)(B - A) - 2\left(\frac{B + E}{2} - \frac{A + F}{2}\right)(E - Q)$$
$$+ (F - B)(C - E) = 0,$$

$$(B + E - A - A)(Q - P) + [(P - C)(P - (A + C - B))$$
$$- (P - (B + E - A))(P - Q)]$$
$$+ (P - C)(C - B) + (A - P)(Q - C) = 0.$$

Example 45. In Figure 4.85, in $\triangle ABC$, where $AB = AC$, D, E, and F are points on lines BC, AB, and AC, respectively, with $DE \parallel AC$ and $DF \parallel AB$. Let M be the midpoint of arc $\widehat{BC}$. Prove that $MD \perp EF$. (2005 Iranian National Team Selection Exam)

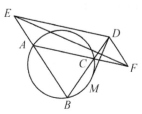

Figure 4.85

Proof. Let the circumcentre O of $\triangle ABC$ be at the origin. Then, $M = -A$:

$$(-A - D)(E - (A + D - E)) - 2((B - A)(B - D) - (B - C)(B - E))$$
$$+ 4\frac{B+C}{2}\left(B - \frac{D+C}{2}\right) + 2\left(\frac{B+D}{2} - E\right)(C - D)$$
$$- 4\frac{B+A}{2}(B - E) + (C^2 - A^2) = 0.$$

Solving geometry problems can sometimes require extremely clever insights that are challenging to come up with. When there seems to be no obvious breakthrough, one can feel somewhat helpless. However, the method of using identities offers a structured approach. It involves representing geometric relationships in the form of vector equations and solving them using the method of undetermined coefficients. This way, geometric problems are transformed into algebraic problems, eliminating the need for sudden flashes of inspiration, and instead, one can proceed systematically by solving equations step by step.

However, it is important to note that every method has its limitations, and the method of identity is no exception. Some geometric relationships are difficult to express using vector equations, especially those involving the addition of two line segments equalling a third line segment, or problems related to finding intersection points or angle relationships. Establishing identities in such cases can be quite challenging. Additionally, while solutions based on identities may appear concise, they often hide complex underlying calculations, requiring diligent practice to become proficient.

It is worth mentioning that the use of identities is not limited to point geometry alone. Identities can be established for angles, lengths, areas, and even coordinates, and these identities can be employed to further investigate and prove various geometric propositions. This opens up avenues for continued research and exploration.

Exercise 4.2

1. In Figure 4.86, the diagonals of quadrilateral $ABCD$ intersect at point P, and M and N are the midpoints of AD and BC, respectively. The circumcentres of $\triangle APB$ and $\triangle CPD$ are denoted as O_1 and O_2, respectively. Prove that $MN \perp O_1O_2 \iff AC = BD$.

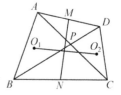

Figure 4.86

2. In Figure 4.87, let O be the circumcentre of $\triangle ABC$. Points D, E, and F lie on sides BC, CA, and AB, respectively, such that $DE \perp CO$ and $DF \perp BO$. K is the circumcentre of $\triangle AFE$. Prove that $DK \perp BC$. (2012 European Girls' Mathematical Olympiad)

3. In Figure 4.88, it is known that CD is a chord of circle O and parallel to the diameter AB. Circle P has its centre on AB and is tangent to lines AA_1, BB_1, CC_1, and DD_1, where A_1, B_1, C_1, and D_1 are the points of tangency. Prove that $AA_1^2 + BB_1^2 = CC_1^2 + DD_1^2$. (Problem Solving 2091 in *Mathematical Bulletin*)

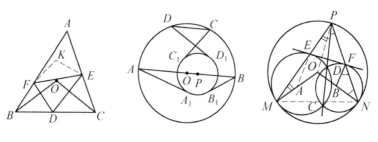

Figure 4.87   Figure 4.88   Figure 4.89

4. In Figure 4.89, circles A and B intersect at points C and D, and they are both internally tangent to circle O with tangent points M and N. The ray CD intersects circle O at point P. Line PM intersects circle A at E, and line PN intersects circle B at F. Prove that EF is a common tangent to circles A and B. (Problem Solving 1222 in *Mathematical Bulletin*)

5. In Figure 4.90, consider hexagon $ABCDEF$, where $AB \parallel DE$, $BC \parallel EF$, and $CD \parallel FA$. Let BD and AE, AC and DF, and CE and BF intersect at points M, N, and K, respectively. Prove that the perpendiculars drawn from M, N, and K to AB, CD, and EF, respectively, intersect at a common point.

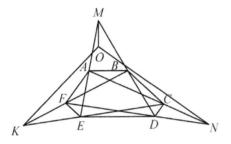

Figure 4.90   Figure 4.91

6. In Figure 4.91, square $ABCD$ has a side length of 1, and points E, H, F, and G lie on sides AB, BC, CD, and DA, respectively, such that $BE + BH + DF + DG = 2$. Prove that $EF \perp GH$.

7. In Figure 4.92, points D, E, and F lie on sides BC, BA, and AC, respectively, and points A, B, I, and C are concyclic. If $ED = EB$, $FD = FC$, and $AI \perp BC$, prove that $ID \perp EF$.

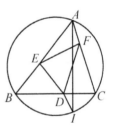

Figure 4.92

8. In Figure 4.93, $ABCD$ is a cyclic quadrilateral with AB as a diameter. $MA \perp AB$, $NB \perp AB$, and points K, E, and L are the midpoints of AD, DC, and CB, respectively. Prove that $MK \perp AE$ and $NL \perp BE$, and deduce that $MN \parallel DC$.

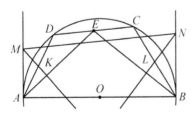

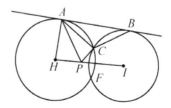

Figure 4.93   Figure 4.94

9. Circles H and I intersect at points C and F. Line AB is a common tangent. See Figure 4.94, and prove that $AC \perp PC \iff AP \perp BC$.

10. Consider quadrilateral $ABCD$. AC intersects BD at E. $FA \perp AC$, and $FD \perp DB$. See Figure 4.95, and prove that $EF \perp BC \iff$ points A, B, C, and D are concyclic.

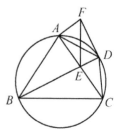

Figure 4.95

11. In Figure 4.96, let the centre of circle A be on circle O, point P be on circle O, PE be the tangent to circle A, and ED be perpendicular to PA at D. Let M and N be two intersection points of the circles. Prove that points M, D, and N are collinear.

12. In Figure 4.97, in $\triangle ABC$, where $\angle C = 90°$ and CD is an altitude. A circle with centre A and radius AC is drawn, and point K is on circle A. A circle O with diameter AE is drawn. Prove that $KD \perp AE$.

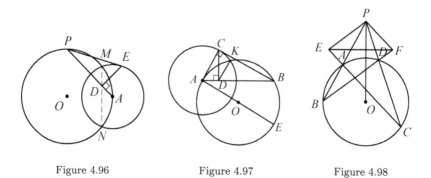

Figure 4.96 Figure 4.97 Figure 4.98

13. In Figure 4.98, let A, B, C, and D be any four points on circle O. Extend BA and CD to meet at point P. Draw parallel lines to BD and AC through P, intersecting the extensions of CA and BD at E and F, respectively. Prove that $OP \perp EF$.

14. In $\triangle ABC$ with $\angle C = 90°$, points D and E lie on side AB, and points P and Q lie on sides AC and BC, respectively. $DE = \frac{1}{2}AB$, $PD \perp AB$, and $QE \perp AB$. Prove that $AP^2 + BQ^2 = CP^2 + CQ^2$.

15. In Figure 4.99, in equilateral $\triangle ABC$ with $\angle D = 60°$, $BE \parallel AD$. Prove that $BE = CD$.

16. In Figure 4.100, in convex quadrilateral $ABCD$ with $\angle A = \angle C$, points M and N are taken on sides AB and BC, respectively, such that $MN \parallel AD$ and $MN = 2AD$. K is the midpoint of MN, and H is the orthocentre of $\triangle ABC$. Prove that $KH \perp CD$. (2018 44th Russian Mathematical Olympiad)

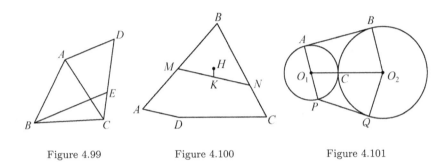

Figure 4.99　　　　　Figure 4.100　　　　　Figure 4.101

17. In Figure 4.101, two circles are externally tangent at point C, and AB is their common external tangent line. Points A and B are the points of tangency, and AP is the diameter of circle O_1. Line PQ is tangent to circle O_2 at Q. Prove that $AP = PQ$. (2006 British Mathematical Olympiad)

18. In Figure 4.102, M is the midpoint of side BC of $\triangle ABC$. The circumcircle of $\triangle ABM$ intersects side AC at E, and the circumcircle of $\triangle AMC$ intersects side AB at F. O is the circumcentre of $\triangle AEF$. Prove that $OB = OC$. (2017 Belarus Mathematical Olympiad)

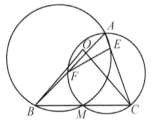

Figure 4.102

19. In Figure 4.103, H is the orthocentre of $\triangle ABC$, and points E and F lie on AC and AB, respectively. Points B, C, E, and F are concyclic. Lines BE and CF intersect at point D. The midpoints of BE and CF are M and N, respectively. Prove that $DH \perp MN$.

20. In Figure 4.104, in $\triangle ABC$, $AD \perp BC$ and $BE \perp CA$. Lines AD and BE intersect at point H. Point P is the midpoint of side AB, and line CQ is drawn through point C perpendicular to PH, where Q is the foot of the altitude. Prove that $PE^2 = PH \cdot PQ$.

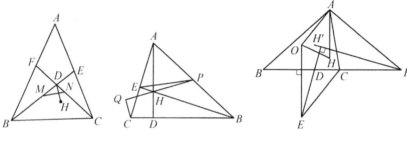

Figure 4.103   Figure 4.104   Figure 4.105

21. In Figure 4.105, O and H are the circumcentre and orthocentre of $\triangle ABC$, respectively. Point H' is the orthocentre of $\triangle AOH$. Point D lies on BC, and $BD = 2DC$. Point E lies on the perpendicular bisector of BC such that $CE \parallel AO$, and $AO \perp AF$. Point F lies on line BC. Prove that $DE \perp H'F$.

22. In Figure 4.106, in $\triangle ABC$, H is the orthocentre. Using H as the centre, draw circles with radius HA, intersecting lines AB and AC at points M and N, respectively. Prove that the five points B, M, H, C, and N are concyclic, and the radius of this circle is equal to the radius of the circumcircle of $\triangle ABC$.

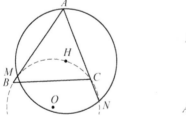
Figure 4.106

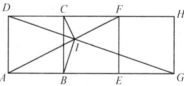
Figure 4.107

23. In Figure 4.107, squares $ABCD$, $BEFC$, and $EGHF$ are given. DG intersects AF at I. Prove that $CI \perp AF$, $DI \perp BI$, and the points A, B, I, and D are concyclic.

24. In Figure 4.108, squares $ABCD$ and $DEFG$ are given. E lies on CD, H lies on BC, and G, E, and H are collinear. Through H, draw $HK \perp AE$, where K lies on DG. Prove that $HA = HK$.
25. In Figure 4.109, In the right trapezium $ABCD$, E and F are on the oblique side AB, and $AD^2 = AE \cdot AF$, $BC^2 = BE \cdot BF$. M is the midpoint of CD. Prove that $ME = MF$.

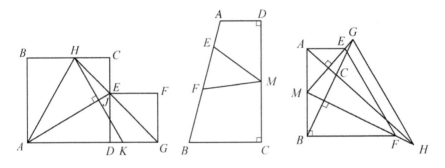

Figure 4.108   Figure 4.109   Figure 4.110

26. In Figure 4.110, in trapezium $ABFE$ with $\angle A = \angle B = 90°$, take the midpoint M of AB. Draw perpendiculars ME and MF from A and B, meeting lines MF and ME at H and G, respectively. Prove that $GH \parallel EF$.
27. In Figure 4.111, consider that P is an external point to circle O and PM is tangent to circle O at point M. Draw lines through P that intersect circle O at points A and B. On segment AB, there exists a point Q such that $\frac{AQ}{QB} = \frac{AP}{BP}$. Prove that $MQ \perp OP$.

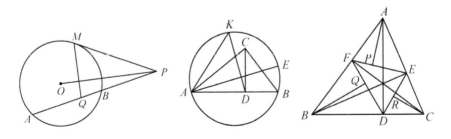

Figure 4.111   Figure 4.112   Figure 4.113

28. In Figure 4.112, CD is the altitude of Rt$\triangle ABC$. Using A as the centre, draw a circle with a radius AC. Point K lies on the circle. Draw a circle O passing points A, K, and B with diameter AE, and prove that $KD \perp AE$.

29. In Figure 4.113, in $\triangle ABC$, AD, BE, and CF are altitudes. $AP \perp EF$, $BQ \perp DE$, and $CR \perp DE$. Prove that lines AP, BQ, and CR are concurrent.

30. In Figure 4.114, two circles are externally tangent at point P. Two mutually perpendicular secants, APA_1 and BPB_1, are drawn through P. If the diameters of the two circles are d_1 and d_2, prove that $AA_1^2 + BB_1^2 = (d_1 + d_2)^2$.

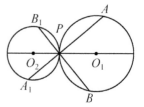

Figure 4.114

31. In Figure 4.115, F lies on the extension of side BC of parallelogram $ABCD$. Draw a circle through points A and E, intersecting AB at E and intersecting DF at G. O is the circumcentre of $\triangle CDF$. Prove that $OD \perp EG$.

32. In Figure 4.116, in $\triangle ABC$, $AB = AC$. A circle with diameter AB intersects BC at D and intersects AC at E. Connect BE, and extend it to intersect AD at F. Let L be the midpoint of AF. Draw EG perpendicular to BC at G. Prove that $CD^2 = EG \cdot DL$.

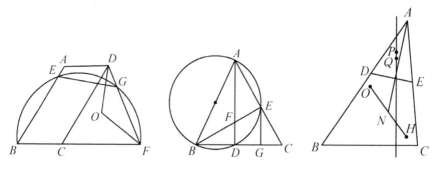

Figure 4.115 Figure 4.116 Figure 4.117

33. In Figure 4.117, O and H are the centroid and orthocentre of $\triangle ABC$, respectively. N is the midpoint of OH. Points D and E lie on AB and

AC, respectively. Points P and Q are the circumcentre and orthocentre of $\triangle ADE$, respectively. Prove that $PQ \perp BC \iff AN \perp DE$.

34. In Figure 4.118, O and H are the circumcentre and orthocentre of $\triangle ABC$, respectively, and N is the midpoint of OH. Point K is the symmetric point of N about BC, and point X lies on BC such that $NA \perp NX$. Prove that $KX \perp KO$.

35. In Figure 4.119, it is known that H is the orthocentre of $\triangle ABC$, E lies on the midline AD, and $EH \perp AD$. Point F is on EB, and $AF \perp AB$. Prove that $AF = EF$.

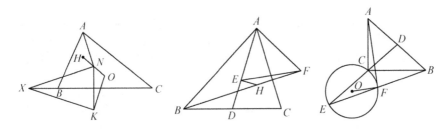

Figure 4.118　　　　　Figure 4.119　　　　　Figure 4.120

36. In Figure 4.120, E is a point on the median CD of Rt$\triangle ACB$ such that $2EC = EB$. F is the middle point of BE. Prove that AF is a tangent to the circumcircle of $\triangle CEF$.

37. In Figure 4.121, $BD \perp AC$ at D, $CE \perp AB$ at E, P is the centre of nine-point circle of $\triangle ADE$, and O and H are the circumcentre and orthocentre of $\triangle ABC$, respectively. Q lies on the line OH, and $HQ = 3OQ$. Prove that $PQ \perp DE$.

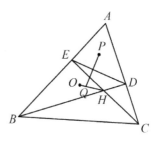

Figure 4.121

38. In Figure 4.122, M and N are the midpoints of sides AB and AC of acute $\triangle ABC$, respectively. D is the foot of an altitude from A to BC. The circumscribed circles of $\triangle ABM$ and $\triangle ACM$ intersect at point P (where $P \neq D$). Prove that PD bisects MN. (2007 33rd Russian Mathematical Olympiad)

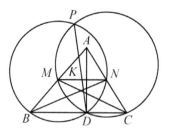 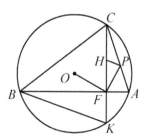

Figure 4.122　　　　　　　　　　Figure 4.123

39. In Figure 4.123, in acute $\triangle ABC$ with $BC > AC$, O is the circumcentre and H is the orthocentre. CF is an altitude, and a line through F is drawn perpendicular to OF, intersecting side CA at P. Prove that $\angle FHP = \angle BAC$.
40. In Figure 4.124, circles O_1 and O_2 intersect at points A and B. A tangent line through A intersects O_1 at C, and the line CB intersects O_2 at D. Line DA intersects O_1 at E. Prove that $CE^2 + DA \cdot DE = CD^2$.
41. In Figure 4.125, BB' and CC' are altitudes of acute $\triangle ABC$. Two circles, P' and Q', pass through A and C' and are tangent to line BC at P and Q, respectively. Prove that A, B', P, and Q are concyclic.

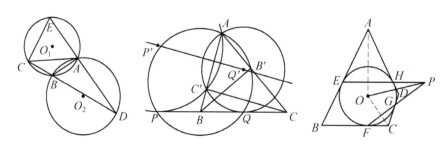

Figure 4.124　　　　　　Figure 4.125　　　　　　Figure 4.126

42. In Figure 4.126, quadrilateral $ABCD$ is circumscribed around circle O. The tangents points on AB, BC, CD, and DA are E, F, G, and H, respectively. Line HE intersects FG at point P. Prove that $OP \perp AC$.

Identity-Based Method 2: Undetermined Coefficients   143

43. In Figure 4.127, point P is outside circle O, and tangents PA and PB are drawn from P. The line through P intersects circle O at points M and N and intersects line AB at point Q. Prove that P, Q, M, and N form a harmonic quadruple.

44. A line DE is on the plane of $\triangle ABC$. Let $AA_1 \perp DE$ at A_1, and $BB_1 \perp DE$ at B_1. If $A_1 B = A_1 C$ and $B_1 A = B_1 C$, prove that $C_1 A = C_1 B$.

45. In Figure 4.128, in $\triangle ABC$, with $BA \perp BC$, BD is the altitude, and BE is the median. A line drawn from D perpendicular to BE intersects line BA at F and line BC at G. Prove that $DF = DG$.

46. Define the centroid of a quadrilateral as the intersecting point of two lines connecting the midpoints of opposite sides. Let $ABCDEF$ be a hexagon inscribed in a circle with centre O, such that $AD = BE = CF$. Let X, Y, and Z be the centroids of quadrilaterals $ABDE$, $BCEF$, and $CDFA$, respectively. Prove that O is the orthocentre of $\triangle XYZ$.

47. In Figure 4.129, in acute $\triangle ABC$, AH is an altitude, and AM is a median. Points X and Y lie on lines AB and AC, respectively, such that $AX = XC$ and $AY = YB$. Prove that the midpoint of XY is equidistant from points H and M.

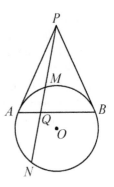

Figure 4.127

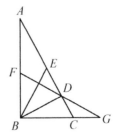

Figure 4.128

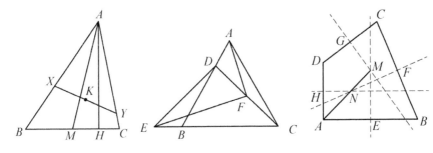

Figure 4.129 Figure 4.130 Figure 4.131

48. In Figure 4.130, in equilateral $\triangle ABC$, point D lies on side AB, and point E lies on the extension of side CB. It is given that $AD = BE$, and point F is the midpoint of CD. Prove that $AF \perp FE$ and $DC = DE$.
49. In Figure 4.131, in quadrilateral $ABCD$, where $AB \perp AD$, $AC = BD$, and the midperpendiculars of AD and BC intersect at N. The midperpendiculars of AB and CD intersect at M. Prove that points A, M, and N are collinear.

Chapter 5

Assorted Problems

5.1 Introducing Parameters

In general, plane geometry is considered simpler than solid geometry, and low-dimensional geometry is simpler than high-dimensional geometry. Therefore, when studying geometry, it is common to start with plane geometry before moving on to three-dimensional space. In recent years, in order to reduce the level of difficulty in learning, large portions of solid geometry content have been omitted. As a result, when elementary geometry is mentioned, it is almost always assumed to be plane geometry.

The vector method for studying geometry is essentially not restricted by the dimension of space. Point geometry inherits this advantage from vector geometry. If a problem involves $n+1$ points, it is assumed to be in an n-dimensional space, and then the dimension is reduced based on the relationships between the points. For example, with three points, A, B, and C, it is initially considered a two-dimensional space. However, if it is known that C is the midpoint of AB, then the three points together generate a one-dimensional space.

Example 1. For four points, A, B, C, and H, if $AH \perp BC$ and $BH \perp CA$, prove that $CH \perp AB$.

Proof. We have the identity
$$(A-B)(H-C) + (B-C)(H-A) + (C-A)(H-B) = 0.$$

Note. This identity appears to prove the orthocentre theorem for triangles. However, it goes beyond that. Since there are no constraints on the

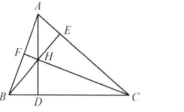
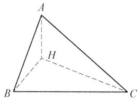

Figure 5.1 Figure 5.2

position of point H, it could be on the plane of ABC (Figure 5.1) or outside of it (Figure 5.2). This is one of the advantages of the point geometry identity method: it effortlessly extends propositions that hold in lower dimensions to higher dimensions. Conversely, many high-dimensional problems can also be treated just like lower-dimensional ones. Other methods, after successfully solving a problem in lower dimensions, would require a completely new approach to determine if the proposition extends to higher dimensions, which can be challenging, time-consuming, and labor-intensive.

If we predefine H to lie on a plane and let $H = \frac{xA+yB+zC}{x+y+z}$, the equation is evidently valid. However, the expression becomes more complex, and the range of validity narrows. The value of the identity decreases. This is also why we do not initially restrict the range of points, in the hope of studying geometric relationships over the broadest possible range. Of course, some propositions are only valid in low-dimensional spaces and fail to hold in higher-dimensional spaces. In such cases, we must gradually reduce dimensions to find the maximum dimension for which the geometric relationship holds.

Example 2. In Figure 5.3, where AA_1, BB_1, and CC_1 intersect at point H, and it is known that the points B, A_1, H, and C_1 form a cyclic quadrilateral, and similarly, the points C, A_1, H, and B_1 also form a cyclic quadrilateral. Prove that the points A, C_1, H, and B_1 also form a cyclic quadrilateral. (1987 Soviet Union National Team Winter Training Camp Test)

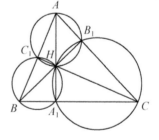

Figure 5.3

Analysis. In this problem, if point H is not on the plane of $\triangle ABC$, then the points of intersection A_1, B_1, and C_1 would not exist, and the concept of a cyclic quadrilateral would not apply. Therefore, constraining H to lie on the plane of $\triangle ABC$ and expressing H in terms of A, B, C, and relevant parameters are necessary.

Proof. Let $H = \frac{xA+yB+zC}{x+y+z}$. Then, we can determine A_1, B_1, and C_1 as follows: $A_1 = \frac{yB+zC}{y+z}$, $B_1 = \frac{xA+zC}{x+z}$, and $C_1 = \frac{xA+yB}{x+y}$. We can then generate the following identity:

$$[(B - C_1)(B - A) - (B - H)(B - B_1)]$$
$$+ [(A - C_1)(A - B) - (A - H)(A - A_1)]$$
$$+ \frac{z}{y}[(A - C)(A - B_1) - (A - H)(A - A_1)] = 0.$$

When the solution using the direct application of identity methods fails, it is necessary to attempt to represent certain points by combining them with reference points (usually the three points introduced initially) using parameters. These are referred to as parameter points.

Example 3. In Figure 5.4. Let the extensions of the opposite sides AB and DC of quadrilateral $ABCD$ intersect at point E, and the extensions of the opposite sides AD and BC intersect at point F. Then, the midpoints of AC (denoted as L), BD (denoted as M), and EF (denoted as N) are collinear. This line is known as the Gauss line.

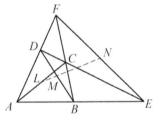

Figure 5.4

Proof. Let $D = \frac{xA+yB+zC}{x+y+z}$, $E = \frac{xA+yB}{x+y}$, $F = \frac{yB+zC}{y+z}$, $L = \frac{A+C}{2}$, $M = \frac{B + \frac{xA+yB+zC}{x+y+z}}{2}$, and $N = \frac{\frac{yB+zC}{y+z} + \frac{xA+yB}{x+y}}{2}$. Now, consider the parameter t such that

$$t\frac{A+C}{2} + (1-t)\frac{B + \frac{xA+yB+zC}{x+y+z}}{2} = \frac{\frac{yB+zC}{y+z} + \frac{xA+yB}{x+y}}{2}.$$

Solving for t, we get $t = \frac{xz}{(x+y)(y+z)}$. Therefore, points L, M, and N are collinear.

Note. By equating the coefficients of A, B, and C on both sides of the equation, three equations can be derived. It is only necessary to solve the simplest among them and then substitute the solutions into the other two equations for verification. Thus, a geometric theorem is equivalent to an algebraic identity:

$$\frac{xz}{(x+y)(y+z)} \cdot \frac{A+C}{2} + \left(1 - \frac{xz}{(x+y)(y+z)}\right) \cdot \frac{B + \frac{xA+yB+zC}{x+y+z}}{2}$$

$$= \frac{\frac{yB+zC}{y+z} + \frac{xA+yB}{x+y}}{2}.$$

Example 4. In Figure 5.5, $\odot I$ is the incircle of $\triangle ABC$. D, E, and F are the points of tangency on BC, CA, and AB, respectively. DD_1, EE_1, and FF_1 are all diameters of $\odot I$. Prove that lines AD_1, BE_1, and CF_1 are concurrent. (Problem Solving 1396 in *Mathematical Bulletin*)

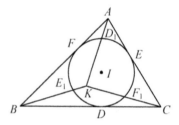

Figure 5.5

Proof.

$$\frac{(-a+b+c)A + (a-b+c)B + (a+b-c)C}{a+b+c}$$

$$= \frac{-3a+b+c}{-a+b+c} A + \left(1 - \frac{-3a+b+c}{-a+b+c}\right)$$

$$\cdot \left(2 \frac{aA+bB+cC}{a+b+c} - \frac{\frac{a+b-c}{2}B + \frac{a-b+c}{2}C}{a}\right)$$

$$= \frac{a-3b+c}{a-b+c} B + \left(1 - \frac{a-3b+c}{a-b+c}\right)$$

$$\cdot \left(2 \frac{aA+bB+cC}{a+b+c} - \frac{\frac{-a+b+c}{2}C + \frac{a+b-c}{2}A}{b}\right)$$

$$= \frac{a+b-3c}{a+b-c} C + \left(1 - \frac{a+b-3c}{a+b-c}\right)$$

$$\cdot \left(2 \frac{aA+bB+cC}{a+b+c} - \frac{\frac{a-b+c}{2}A + \frac{-a+b+c}{2}B}{c}\right).$$

Note. Here, $I = \frac{aA+bB+cC}{a+b+c}$, $D = \frac{\frac{a+b-c}{2}B + \frac{a-b+c}{2}C}{a}$, $D_1 = 2\frac{aA+bB+cC}{a+b+c} - \frac{\frac{a+b-c}{2}B + \frac{a-b+c}{2}C}{a}$, and so on.

Assorted Problems   149

Example 5. In Figure 5.6, in $\triangle ABC$, the incircle touches BC and CA at D and E, respectively. On the extension of BA, a point F is taken such that $AF = CD$. Prove that D, E, and F are collinear $\iff$ $AB \perp AC$.

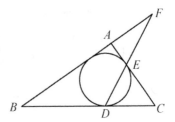

Figure 5.6

Proof. Let $A = 0$, then

$$t \frac{-\frac{a+b-c}{2}B + \frac{a+b+c}{2}A}{c} + (1-t)\frac{\frac{a-b+c}{2}C + \frac{a+b-c}{2}B}{a}$$

$$- \frac{\frac{a+b-c}{2}A + \frac{-a+b+c}{2}C}{b}$$

$$= -\frac{(a+b-c)(-c+ct+at)}{2ac}B + \frac{(a^2+b(b-c)(t-1)-a(c+bt))}{2ab}C.$$

Solving the equation $-\frac{(a+b-c)(-c+ct+at)}{2ac} = 0$ gives $t = \frac{c}{a+c}$. Thus, we obtain the identity

$$\frac{c}{a+c} \cdot \frac{-\frac{a+b-c}{2}B + \frac{a+b+c}{2}A}{c} + \left(1 - \frac{c}{a+c}\right)\frac{\frac{a-b+c}{2}C + \frac{a+b-c}{2}B}{a}$$

$$- \frac{\frac{a+b-c}{2}A + \frac{-a+b+c}{2}C}{b} = \frac{a^2-b^2-c^2}{2b(a+c)}C.$$

Therefore, points D, E, and F are collinear if and only if $AB \perp AC$.

Example 6. In $\triangle ABC$, BD and CE are angle bisectors. Prove that $AB = AC \iff BD = CE$ (Stanley–Lehmer theorem).

Proof. Let $A = 0$, then

$$\left(\frac{aA+bB}{a+b} - C\right)^2 - \left(\frac{aA+cC}{a+c} - B\right)^2$$

$$= \left(-1 + \frac{b^2}{(a+b)^2}\right)B^2 + \left(-\frac{2b}{a+b} + \frac{2c}{a+c}\right)BC + \left(1 - \frac{c^2}{(a+c)^2}\right)C^2$$

$$= \left(-1 + \frac{b^2}{(a+b)^2}\right)c^2 + \left(-\frac{2b}{a+b} + \frac{2c}{a+c}\right)\frac{b^2+c^2-a^2}{2}$$

$$+ \left(1 - \frac{c^2}{(a+c)^2}\right)b^2$$

$$= \frac{a(b-c)(a+b+c)(a^3+a^2b+a^2c+3abc+b^2c+bc^2)}{(a+b)^2(a+c)^2}.$$

To rewrite in the form of an identity, let $A = 0$. Then,

$$\left[\left(\frac{aA+bB}{a+b}-C\right)^2 - \left(\frac{aA+cC}{a+c}-B\right)^2\right.$$

$$-\frac{a(b-c)(a+b+c)(a^3+a^2b+a^2c+3abc+b^2c+bc^2)}{(a+b)^2(a+c)^2}$$

$$-\left(-1+\frac{b^2}{(a+b)^2}\right)(B^2-c^2) - \left(1-\frac{c^2}{(a+c)^2}\right)(C^2-b^2)$$

$$+\left(-\frac{2b}{a+b}+\frac{2c}{a+c}\right)\left(\frac{b^2+c^2-a^2}{2}-BC\right) = 0.$$

Example 7. In Figure 5.7, consider quadrilateral $ABCD$ inscribed in a circle. The extensions of BA and CD intersect at point R, and the extensions of AD and BC intersect at point P. AC intersects BD at point Q. $\angle A$, $\angle B$, and $\angle C$ refer to the interior angles of $\triangle ABC$. Prove that

$$\frac{\cos A}{AP} + \frac{\cos C}{CR} = \frac{\cos B}{BQ}.$$

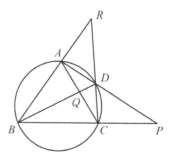

Figure 5.7

Proof. Let $D = \frac{xA+yB+zC}{x+y+z}$, $P = \frac{yB+zC}{y+z}$, $R = \frac{xA+yB}{x+y}$, $Q = \frac{xA+zC}{x+z}$,

$$(Q-B)(R-C) + (Q-B)(P-A) + (P-A)(R-C)$$
$$-\frac{(x+y+z)^2}{(x+z)(y+z)}[(R-A)(R-B)-(R-D)(R-C)] = 0.$$

Note. $\overrightarrow{QB} \cdot \overrightarrow{RC} + \overrightarrow{QB} \cdot \overrightarrow{PA} + \overrightarrow{PA} \cdot \overrightarrow{RC} = 0$ is equivalent to $\frac{\cos A}{AP} + \frac{\cos C}{CR} = \frac{\cos B}{BQ}$. It is also evident that proving $\frac{\cos A}{AP} + \frac{\cos C}{CR} = \frac{\cos B}{BQ}$ implies that points A, B, C, and D are concyclic. Readers can attempt to set $R=0$, $A=tB$, and $D=sC$ to establish the identity and reduce the computational effort.

Exercise 5.1

1. In Figure 5.8, if four vertices of a pentagon have altitudes drawn to the opposite sides that intersect at one point, then the line connecting the remaining vertex to this intersection point is also perpendicular to

the opposite side. In pentagon $ABCDE$, where AF is perpendicular to CD, BG is perpendicular to DE, $CH \perp AE$, and $DI \perp AB$. AF, BG, CH, and DI intersect at point O. Prove that $EO \perp BC$.

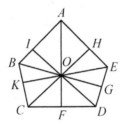

Figure 5.8

2. In Figure 5.9, In $\triangle ABC$, O is a free point. Construct parallelograms $OBXC$, $OCYA$, and $OAZB$. Draw a line through point X perpendicular to BC, a line through point Y perpendicular to CA, and a line through point Z perpendicular to AB. Prove that the three perpendicular lines intersect at one point, K.

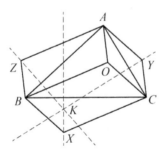

Figure 5.9

5.2 Introducing Complex Numbers

To represent the positions of points more precisely and to deal with some geometric problems involving angles, it is necessary to introduce the complex number multiplication for points in the ASA formula. In $\triangle ABC$, if one side AB and two angles $\alpha = \angle CAB$ and $\beta = \angle CBA$ are known, and when the vertices A, B, and C rotate anticlockwise, the following ASA

formula holds:
$$C = A + \frac{e^{i\alpha}\sin\beta}{\sin(\alpha+\beta)}(B-A) = A + \frac{1-e^{-2i\beta}}{1-e^{-2i(\alpha+\beta)}}(B-A).$$

Example 1. In $\triangle ABC$, as shown in Figure 5.10, equilateral triangles $\triangle BCF$, $\triangle CAD$, and $\triangle BAE$ are constructed externally with sides of the triangle as their side lengths. Here, O_1, O_2, and O_3 are the centroids of $\triangle BCF$, $\triangle CAD$, and $\triangle BAE$, respectively. Prove that $\triangle O_1O_2O_3$ is an equilateral triangle. (Napoleon's theorem)

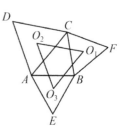

Figure 5.10

Analysis. Let $T = \cos\frac{\pi}{3} + i\sin\frac{\pi}{3}$. The given conditions can be expressed as

$$(D-A)-(C-A)T = 0, \quad (F-C)-(B-C)T = 0, \quad (B-A)-(E-A)T = 0.$$

To prove the conclusion,
$$\left(\frac{A+C+D}{3} - \frac{A+B+E}{3}\right) - \left(\frac{F+B+C}{3} - \frac{A+B+E}{3}\right)T = 0.$$

We introduce complex numbers to deal with the angle issues.

Proof. Assume
$$\left[\left(\frac{A+C+D}{3} - \frac{A+B+E}{3}\right) - \left(\frac{F+B+C}{3} - \frac{A+B+E}{3}\right)T\right]$$
$$+ k_1[(D-A)-(C-A)T] + k_2[(F-C)-(B-C)T]$$
$$+ k_3[(B-A)-(E-A)T] = 0.$$

Expanding by point letters yields
$$\frac{1}{6}A(1+\sqrt{3}i - 3k_1 + 3\sqrt{3}k_1 i - 3k_3 + 3\sqrt{3}k_3 i)$$
$$+ \frac{1}{6}B(-2 - 3k_2 - 3\sqrt{3}k_2 i + 6k_3)$$
$$+ \frac{1}{6}C(1-\sqrt{3}i - 3k_1 - 3\sqrt{3}k_1 i - 3k_2 + 3\sqrt{3}k_2 i)$$
$$+ \frac{1}{3}D(1+3k_1)$$
$$- \frac{1}{6}iE(-i-\sqrt{3} - 3k_3 i + 3\sqrt{3}k_3)$$
$$+ \frac{1}{6}F(-1-\sqrt{3}i + 6k_2) = 0.$$

Solving the equations,

$$(1+\sqrt{3}\mathrm{i}-3k_1+3\sqrt{3}k_1\mathrm{i}-3k_3+3\sqrt{3}k_3\mathrm{i})$$
$$=(-2-3k_2-3\sqrt{3}k_2\mathrm{i}+6k_3)$$
$$=(1-\sqrt{3}\mathrm{i}-3k_1-3\sqrt{3}k_1\mathrm{i}-3k_2+3\sqrt{3}k_2\mathrm{i})$$
$$=(1+3k_1)$$
$$=(-\mathrm{i}-\sqrt{3}-3k_3\mathrm{i}+3\sqrt{3}k_3)$$
$$=(-1-\sqrt{3}\mathrm{i}+6k_2)$$
$$=0.$$

We obtain $k_1 = -\frac{1}{3}$, $k_2 = \frac{1+\sqrt{3}\mathrm{i}}{6}$, and $k_3 = \frac{1+\sqrt{3}\mathrm{i}}{6}$. Thus, we have the identity

$$3\left[\left(\frac{A+C+D}{3}-\frac{A+B+E}{3}\right)-\left(\frac{F+B+C}{3}-\frac{A+B+E}{3}\right)T\right]$$
$$-[(D-A)-(C-A)T]+\frac{1+\sqrt{3}\mathrm{i}}{2}[(F-C)-(B-C)T]$$
$$+\frac{1+\sqrt{3}\mathrm{i}}{2}[(B-A)-(E-A)T]=0.$$

Note. This approach is slightly more intricate than directly computing D, E, and F from the given conditions and substituting. Since the number of equations exceeds the number of unknowns, we can first solve the simpler equations and then substitute the results into the more complex equations for verification. After obtaining the identity, we can derive a more general conclusion. In $\triangle ABC$, when equilateral triangles $\triangle BCF$, $\triangle CAD$, and $\triangle BAE$ are constructed externally with sides of the triangle as their side lengths, and O_1, O_2, and O_3 are the centroids of $\triangle BCF$, $\triangle CAD$, and $\triangle BAE$, respectively, if any three of $\triangle BCF$, $\triangle CAD$, $\triangle BAE$, and $\triangle O_1 O_2 O_3$ are equilateral triangles, then the remaining one must also be an equilateral triangle.

Consideration. Without introducing complex numbers, how can we prove Napoleon's theorem? In this problem, if we use $AC = CD = DA$, it does not uniquely determine point D because the reflection of D across AC also satisfies the requirements. As a result, D cannot be uniquely determined, and consequently, O_2 cannot be determined either, making it challenging to reach a definite conclusion. Research shows that Napoleon's theorem can be proven without complex numbers, but it requires more inner product relations to determine the position of point D.

Constructing equilateral triangles or squares with specified line segments as sides can be problematic due to the existence of two possibilities (on either side). If these possibilities cannot be effectively distinguished, the solution may be difficult. When the straightforward application of identity-based methods fails, especially when angles are involved, particularly special angles such as 30°, 45°, 60°, and 90°, considering the use of complex numbers can be beneficial. Even if the problem doesn't explicitly involve angles, introducing complex numbers can help pinpoint positions accurately and differentiate between different cases effectively.

Example 2. In Figure 5.10, equilateral triangles $\triangle CBF$, $\triangle CAD$, and $\triangle BAE$ are constructed externally with sides equal to the sides of $\triangle ABC$. Prove that $\triangle ABC$ is an equilateral triangle if and only if $\triangle DEF$ is an equilateral triangle.

Proof. Let $T = \cos\frac{\pi}{3} + i\sin\frac{\pi}{3}$. Then,

$$(D-E) - (F-E)T + \frac{1-\sqrt{3}i}{2}[(C-A)-(B-A)T]$$

$$- [(D-A)-(C-A)T] + \frac{1+\sqrt{3}i}{2}[(F-C)-(B-C)T]$$

$$+ \frac{1+\sqrt{3}i}{2}[(B-A)-(E-A)T] = 0.$$

The identity implies a more general conclusion: in Figure 5.10, when equilateral triangles $\triangle CBF$, $\triangle CAD$ and $\triangle BAE$ are constructed externally with sides equal to the sides of $\triangle ABC$, prove that if four out of $\triangle CBF$, $\triangle CAD$, $\triangle BAE$, $\triangle ABC$, and $\triangle DEF$ are equilateral triangles, then the rest one must also be an equilateral triangle.

Example 3. In Figure 5.11, in $\triangle ABC$ with $\angle C = 90°$, $\angle CAD = 30°$, and $AC = BC = AD$, prove that $CD = BD$.

Analysis. Let $C = 0$. Attempting to express $(D-B)^2 - D^2$ linearly in terms of $A^2 - B^2$, AB, $(A-D)^2 - A^2$, and $\frac{\sqrt{3}A^2}{2} - (A-C)(A-D)$ is not successful.

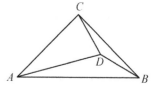

Figure 5.11

This is because in Figure 5.11, point D, which satisfies $DC = DB$, is not unique. The reflection of D about AC also satisfies $AC = AD$ and

∠CAD = 30° but clearly does not satisfy DC = DB. In this case, complex numbers are needed to further determine the position of D. Use $\overrightarrow{AD}e^{\frac{\pi}{6}i} = \overrightarrow{AC}$ and $\overrightarrow{CA}i = \overrightarrow{CB}$ to represent $\overrightarrow{DB}e^{\frac{5\pi}{6}i} = \overrightarrow{DC}$.

Proof.
$$\left[(B-D)\left(\cos\frac{5\pi}{6} + i\sin\frac{5\pi}{6}\right) - (C-D) \right]$$
$$- \frac{(1+\sqrt{3})(1-i)}{2}\left[(D-A)\left(\cos\frac{\pi}{6} + i\sin\frac{\pi}{6}\right) - (C-A) \right]$$
$$+ \frac{i-\sqrt{3}}{2}[(A-C)i - (B-C)] = 0.$$

Note. In this example, it is easy to see that the distance from D to AC is equal to half of AC, so the foot of the perpendicular from D to BC is the midpoint of BC. In comparison, point geometry methods or other methods may not reveal this superiority.

Example 4. In Figure 5.12, in quadrilateral $ABCD$, where $AB \perp AC$, M and N are the midpoints of AD and BC, respectively, equilateral $\triangle CDE$ is outside of quadrilateral $ABCD$, and if $\angle NME = 90°$, find $\angle B$.

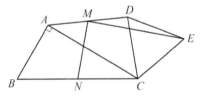

Figure 5.12

Proof. Let
$$[(A-B)z - (A-C)] + k_1\left[\left(\frac{A+D}{2} - \frac{B+C}{2}\right)ti - \left(\frac{A+D}{2} - E\right)\right]$$
$$+ k_2[(D-C)\frac{1+\sqrt{3}i}{2} - (D-E)] = 0,$$
which simplifies to $\frac{1}{2}A(-2 - k_1 + ik_1t + 2z) - \frac{1}{2}iB(k_1t - 2zi) + \frac{1}{2}C(2 - k_2 - i\sqrt{3}k_2 - ik_1t) + \frac{1}{2}iD(ik_1 + ik_2 + \sqrt{3}k_2 + k_1t) + E(k_1 + k_2) = 0$.

Solving the system of equations,
$$-2 - k_1 + ik_1t + 2z = 0,$$
$$k_1t - 2zi = 0,$$
$$2 - k_2 - i\sqrt{3}k_2 - ik_1t = 0,$$
$$ik_1 + ik_2 + \sqrt{3}k_2 + k_1t = 0,$$
$$k_1 + k_2 = 0,$$

we find that $k_1 = -2$, $k_2 = 2$, $t = \sqrt{3}$, and $z = \sqrt{3}i$. Therefore, we have the identity

$$[(A-B)\sqrt{3}i - (A-C)] - 2\left[\left(\frac{A+D}{2} - \frac{B+C}{2}\right)\sqrt{3}i - \left(\frac{A+D}{2} - E\right)\right]$$
$$+ 2\left[(D-C)\frac{1+\sqrt{3}i}{2} - (D-E)\right] = 0.$$

It can be seen that $\angle B = 60°$, $ME \perp MN$, and $ME = \sqrt{3}MN$.

Example 5. In Figure 5.13, in $\triangle ABC$, where $\angle BAC = 90°$, $AC = 2AB$ and point D is the midpoint of AC. A right-angled set square with an acute angle of $45°$ is placed, as shown in the diagram, such that the two endpoints of the hypotenuse coincide with A and D, and BE and EC are connected. Try to guess the quantity and positional relationship of line segments BE and EC, and prove your guess. (2011 Sichuan Secondary School Entrance Examination)

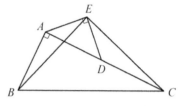

Figure 5.13

Proof.

$$[(E-C) - (E-B)i] + \frac{1}{2}[(C-A) - 2(B-A)i]$$
$$+ \left[\left(\frac{A+C}{2} - E\right) - (A-E)i\right] = 0.$$

This shows that EB and EC are perpendicular and equal.

Example 6. In Figure 5.14, BD and CE are the altitudes of $\triangle ABC$. Point P lies on the extension of BD such that $BP = AC$, and point Q lies on CE, such that $CQ = AB$. Prove that $AP = AQ$ and $AP \perp AQ$.

Proof.

$$[(A-P) - (A-Q)i] - i[(Q-C) - (A-B)i] - [(B-P) - (A-C)i] = 0.$$

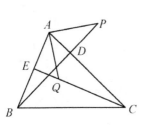
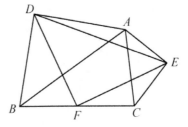

Figure 5.14   Figure 5.15

Example 7. In Figure 5.15, in $\triangle ABC$, construct isosceles right $\triangle DBA$ outward from side AB and $\triangle ACE$ outward from side AC. Let F be the midpoint of BC. Prove that $\triangle DFE$ is an equilateral right triangle. (1996 Irish Mathematical Competition)

Proof.

$$\left[\left(D - \frac{B+C}{2}\right) - \left(E - \frac{B+C}{2}\right)\mathrm{i}\right]$$
$$- \left[D - B - (A-B)\frac{1+\mathrm{i}}{2}\right] + \mathrm{i}\left[E - A - (C-A)\frac{1+\mathrm{i}}{2}\right] = 0.$$

Example 8. In Figure 5.16, in $\triangle ABC$, with $\angle BAC = 90°$, $DB \perp BC$ and $DB = BC$. $EB \perp BA$ and $EB = BA$. The extensions of DA and EC intersect at F. Prove that $AF \perp CF$.

Proof.

$$[(D - A) - \mathrm{i}(C - E)] - [(D - B) - (C - B)\mathrm{i}] + [(A - B) - (E - B)\mathrm{i}] = 0.$$

Example 9. In Figure 5.17, two squares $ABEF$ and $ACGH$ are constructed outward from sides AB and AC of $\triangle ABC$. P is the midpoint of EG. Prove that $BP \perp CP$ and $BP = CP$.

Proof.

$$\left[\left(C - \frac{E+G}{2}\right) - \left(B - \frac{E+G}{2}\right)\mathrm{i}\right] + \frac{1+\mathrm{i}}{2} \cdot [(A-C) - (G-C)\mathrm{i}]$$
$$+ \frac{1-\mathrm{i}}{2}[(E-B) - (A-B)\mathrm{i}] = 0.$$

158 Solving Problems in Point Geometry

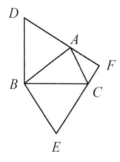

Figure 5.16

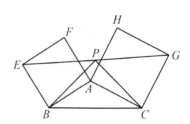

Figure 5.17

Example 10. In Figure 5.18, within square $ABCD$, construct an isosceles $\triangle ABE$ with $\angle EAB = \angle EBA = 15°$. Prove that $\triangle CDE$ is an equilateral triangle.

Proof.

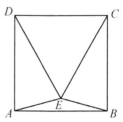

Figure 5.18

$$\left[(E-C)\left(\cos\frac{\pi}{3}+\mathrm{i}\sin\frac{\pi}{3}\right)-(E-(A+C-B))\right]$$
$$+\frac{\sqrt{3}+\mathrm{i}}{2}[\mathrm{i}(C-B)-(A-B)]$$
$$+\frac{-1+\mathrm{i}\sqrt{3}}{2}\left[\frac{\sin\frac{\pi}{12}\left(\cos\frac{\pi}{12}+\mathrm{i}\sin\frac{\pi}{12}\right)}{\sin\frac{\pi}{6}}(B-A)-(E-A)\right]=0.$$

Exercise 5.2

1. In Figure 5.19, in $\mathrm{Rt}\triangle ABC$, where $CA \perp CB$ and $\angle B = 60°$. Construct equilateral triangles $\triangle BAE$ and $\triangle ACD$. Let ED intersect AB at point F. Prove that F is the midpoint of DE.
2. In Figure 5.20, $\triangle ABD$ and $\triangle ACE$ are equilateral triangles. Through point A, draw $AF \parallel CD$ intersecting BC at point F. Prove that $EF \parallel BD$.

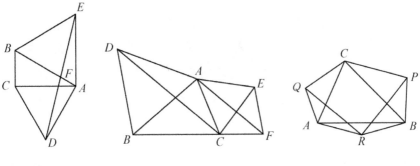

Figure 5.19   Figure 5.20   Figure 5.21

3. In Figure 5.21, on the plane of triangle $\triangle ABC$, in its exterior, we draw triangles $\triangle ABR$, $\triangle BCP$, and $\triangle CAP$ so that $\angle PBC = \angle CAQ = 45°$, $\angle BCP = \angle QCA = 30°$, and $\angle ABR = \angle RAB = 15°$. Prove that $\angle QRP = 90°$ and $QR = RP$. (1975 International Mathematical Olympiad)

4. In Figure 5.22, in quadrilateral $ABCD$ as shown, $BE \perp AB$ and $BE = AB$, $DF \perp AD$ and $DF = AD$, $BG \perp BC$ and $BG = BC$, $DH \perp CD$ and $DH = CD$. Prove that if E, C, and F are collinear, then A, G, and H are also collinear. (Provided by Ye Zhonghao)

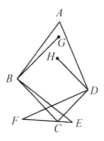

Figure 5.22

5. In Figure 5.23, consider hexagon $ABCDEF$. Using its sides, construct equilateral triangles $\triangle ABC_1$, $\triangle BCD_1$, $\triangle CDE_1$, $\triangle DEF_1$, $\triangle EFA_1$, and $\triangle FAB_1$ outwardly. Prove that if $\triangle B_1D_1F_1$ is equilateral, so is $\triangle A_1C_1E_1$.

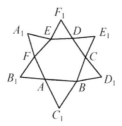

Figure 5.23

6. In Figure 5.24, construct isosceles right triangles $\triangle BAA_1$, $\triangle CBB_1$, $\triangle DCC_1$, and $\triangle ADD_1$ outwardly on the sides of quadrilateral $ABCD$, where M, N, P, and Q are the midpoints of segments A_1B_1, B_1C_1, C_1D_1, and D_1A_1, respectively. Prove that quadrilateral $MNPQ$ is a square.

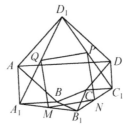

Figure 5.24

7. In Figure 5.25, $ABCD$ is a quadrilateral, and E, F, G, and H are the midpoints of AB, BC, CD, and DA, respectively. Construct isosceles right triangles $\triangle FEA_1$, $\triangle GFB_1$, $\triangle HGC_1$, and $\triangle EHD_1$ outwardly using segments EF, FG, GH, and HE as bases, respectively. Prove that quadrilateral $A_1B_1C_1D_1$ is a square.

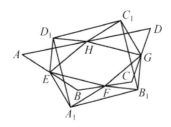

Figure 5.25

5.3 Combined with Other Methods

Point geometry methods are highly effective in many problem-solving scenarios, but they are not universal. Sometimes, when combined with other methods, the range of problems that can be solved expands significantly. In fact, in previous chapters, we have used this method several times. Here, we specifically reintroduce it, hoping to draw everyone's attention.

Example 1. Consider $\triangle ABC$, as shown in Figure 5.26, where $AB = AC$. Point D lies on the base BC, and E is a point on AD such that $\angle ECB = \angle DAC$ and $BD = 2DC$. Prove that $AD \perp BE$.

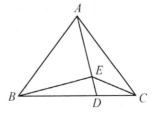

Figure 5.26

Analysis. The point geometry method is adept at exploiting relationships between line segments but is less proficient with angle relationships. To tackle this problem, we need to transform the angle relationship $\angle ECB = \angle DAC$ into a form that can be utilized. We can start by using $\angle ECD = \angle DAC$ and $\angle CDE = \angle ADC$ to deduce that $\triangle DCE \sim \triangle DAC$, leading to the segment relationship $DC^2 = DA \cdot DE$.

Proof. Let $A = 0$. Then,

$$3\left(\frac{B+2C}{3}\right)(B-E) + 3\left[\left(\frac{B+2C}{3} - C\right)^2 - \frac{B+2C}{3}\left(\frac{B+2C}{3} - E\right)\right]$$
$$- (B^2 - C^2) = 0.$$

Example 2. Consider $\triangle ABC$, as shown in Figure 5.27, where $AB = AC$. The three altitudes are AM, BK, and CL. Line ME intersects AB and AC at points E and F, respectively. Prove that $FA \cdot EL + FA \cdot FK = EF^2$.

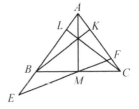

Figure 5.27

Proof. Let $A = 0$, then we have

$$[E(E-L) + F(F-K) - (E-F)^2] + [F(B-E) - E(F-C)]$$
$$+ (L-C)E + (K-B)F = 0.$$

Note. By Menelaus's theorem, we have $\frac{AE}{EB} \cdot \frac{BM}{MC} \cdot \frac{CF}{FA} = 1$, which implies

$$(A - F)(B - E) - (A - E)(F - C) = 0.$$

Therefore, we have shown that $EA \cdot EL + FA \cdot FK = EF^2$.

Example 3. In Figure 5.28, let G be the centroid of $\triangle ABC$ and G_1 be the point symmetric to G about side BC. Prove that the points A, B, G_1, and C are concyclic if and only if $AB^2 + AC^2 = 2BC^2$.

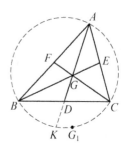

Figure 5.28

Analysis. The direct application of point geometry identity in this problem is cumbersome because it is not straightforward to represent the concyclicity of A, B, G_1, and C. However, if we note the parallelogram $CGBK$, it is easy to see that $\angle CG_1B = \angle CGB = \angle CKB$, implying that K, B, G_1, and C are concyclic. We prove the equivalent statement: K, B, G_1, and C are concyclic if and only if $AB^2 + AC^2 = 2BC^2$. This is evident using identities.

Proof. $(A-B)^2 + (A-C)^2 - 2(B-C)^2 = 6\left[\left(\frac{B+C}{2} - B\right)\left(\frac{B+C}{2} - C\right) - \left(\frac{B+C}{2} - A\right)\left(\frac{B+C}{2} - \left(2\frac{A+B+C}{3} - A\right)\right)\right].$

Example 4. In Figure 5.29, given $\triangle ABC$, where H is the intersection of altitudes AD and BE, P is the midpoint of CH, M is the midpoint of AB, and K is the intersection of lines ED and AB, prove that $MC \perp HK$.
(2010 India National Training Camp Test)

Analysis. This problem can be solved in two steps. First, prove $MP \perp ED$.

Proof. Let $H = 0$, then

$$2\left(\frac{A+B}{2} - \frac{C}{2}\right)(D - E) - D(B - C) - A(D - C) - E(C - A)$$

$$+ B(E - C) - C(A - B) = 0.$$

$$2K\left(C - \frac{B+A}{2}\right) + 2\left(\frac{C}{2} - \frac{B+A}{2}\right)(D - K) + C(A - K)$$

$$- D(D - B) + (D + A)(D - C) = 0.$$

Assorted Problems   163

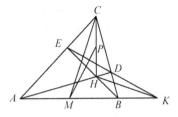 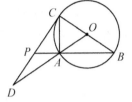

Figure 5.29   Figure 5.30

Example 5. In Figure 5.30, let P be an external point to circle O, and the secant line PAB intersects circle O at points A and B. Line BC is drawn through point B as a diameter, and the extensions of OA and CP intersect at point D, where $DA^2 = DP \cdot DC$. Prove that CD is a tangent to circle O.

Proof. Let $D = 0$, then
$$(C - B)P + (A^2 - PC) - (A - C)(A - P) + [C(P - A) - P(A - B)] = 0.$$

Note. According to Menelaus' theorem, using $\frac{BO}{OC} \cdot \frac{CD}{DP} \cdot \frac{PA}{AB} = 1$, we have $(C - D)(P - A) = (P - D)(A - B)$.

Example 6. In Figure 5.31, let $\triangle ABC$ be given. Extend AC to D such that $CA = CD$. Through D, draw a line intersecting AB at F and BC at E such that points A, B, G, and C are concyclic. O is the circumcentre of $\triangle BDF$. Prove that $OG \perp GA$.

Proof. Let $O = 0$, then $G(E - A) + [(E - A)(E - G) - (E - B)(E - C)] - \frac{B+F}{2}(F - A) + \frac{2C-A+F}{2}(F - E) + \frac{1}{2}[(A - B)(E - F) - (F - B)(2C - A - E)] = 0$.

Note. According to Menelaus' theorem, using $\frac{AB}{BF} \cdot \frac{FE}{ED} \cdot \frac{DC}{CA} = 1$, we have $\overrightarrow{AB} \cdot \overrightarrow{EF} = \overrightarrow{FB} \cdot \overrightarrow{DE}$, which simplifies to $(A - B)(E - F) - (F - B)(2C - A - E) = 0$.

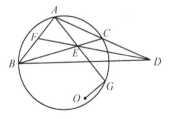 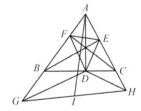

Figure 5.31   Figure 5.32

Example 7. In Figure 5.32, let AD, BE, and CF be the altitudes of $\triangle ABC$. Draw $DG \perp DF$, intersecting line AB at G. Draw $DH \perp DE$, intersecting line AC at H. Point I is the midpoint of GH. Prove that $AI \perp EF$.

Proof.

$$2\left(A - \frac{G+H}{2}\right)(E - F) + (A - G)(F - C) - (A - H)(E - B)$$
$$+ [(E - F)(B - D) - (E - C)(B - G)]$$
$$+ [(F - B)(C - H) - (F - E)(C - D)]$$
$$+ [(A - F)(D - B) - (A - E)(D - F)]$$
$$+ [(A - F)(D - E) - (A - E)(D - C)] = 0.$$

Note. It is easy to prove $\triangle EFC \sim \triangle BGD$, which gives $\frac{EF}{BG} = \frac{EC}{BD}$, so $(E - F)(B - D) = (E - C)(B - G)$.

Example 8. In Figure 5.33, in $\triangle ABC$, squares $ACED$ and $BCFG$ are constructed on sides CA and CB, respectively. Prove that $EF = 2CM$, where M is the midpoint of AB.

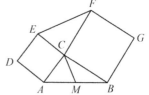

Figure 5.33

Proof. Let $C = 0$, then

$$\left[4\left(\frac{A+B}{2}\right)^2 - (E - F)^2\right] - (A^2 - E^2) - (B^2 - F^2) - 2(AB + EF) = 0.$$

Note. Constructing squares on one side has two possibilities: inside or outside $\triangle ABC$. From the equation, we can see that for $A^2 = E^2$ and $B^2 = F^2$, $4\left(\frac{A+B}{2}\right)^2 = (E - F)^2 \Leftrightarrow AB + EF = 0$, which means that the conclusion is only correct when squares are constructed both outside or inside $\triangle ABC$.

Example 9. In Figure 5.34, in the cyclic quadrilateral $ABCD$ where $BC > AD$ and $CD > AB$, points E and F are on BC and CD, respectively, such that $BE = AD$ and $DF = AB$. M is the midpoint of EF. Prove that $DM \perp BM$.

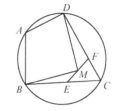

Figure 5.34

Proof.

$$4\left(\frac{E+F}{2} - D\right)\left(\frac{E+F}{2} - B\right) - 2[(B-E)(D-F) + (D-A)(B-A)]$$
$$+ [(A-D)^2 - (B-E)^2] + [(A-B)^2 - (D-F)^2] = 0.$$

From the identity, we can see that the proposition can be expanded to: in the convex quadrilateral $ABCD$, where points E and F are on rays BC and CD, respectively, such that $BE = AD$ and $DF = AB$, M is the midpoint of EF. Prove that the four points A, B, C, D are concyclic if and only if $DM \perp BM$.

Note. There are two possible positions for fixing the length BE on line BC. When paired in all possible combinations, there are four potential configurations, but only two of them are correct.

Example 10. In Figure 5.35, both AB and CD are diameters of $\odot O$, where $\widehat{AC} = 60°$. Choose a point P on the minor arc $\widehat{CB}$. Lines PA and PD intersect CD and AB at M and N, respectively. Prove that $PA \cdot PD = PA \cdot PM + PD \cdot PN$. (Solution to Problem 867 in *Mathematical Bulletin*)

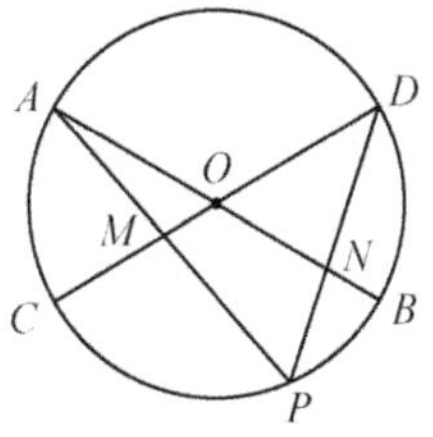

Proof. Let $O = 0$, then

Figure 5.35

$$[2(P-A)(P+C) - (P-A)(P-M) - (P+C)(P-N)]$$
$$+ [A^2 - (A-C)^2] + [A(A-N) - (A-M)(A-P)]$$
$$+ (P-C)(-C-N) - 2\left(C - \frac{A}{2}\right)N = 0.$$

Note. Note that $2\overrightarrow{PA} \cdot \overrightarrow{PD} = |\overrightarrow{PA}| \cdot |\overrightarrow{PD}|$. Also, points O, M, P, and N are concyclic. The signs of $\overrightarrow{PA} \cdot \overrightarrow{PM}$ and $\overrightarrow{PD} \cdot \overrightarrow{PN}$ depend on the position of P and can be different.

Example 11. In Figure 5.36, in $\triangle ABC$, where D and E are the midpoints of AB and AC, respectively, semicircles are constructed outside $\triangle ABC$ with AB and AC as diameters. Parallel lines to AC through D intersect semicircle D at F. Parallel lines to AB through E intersect semicircle E at G. Tangents are drawn from F to semicircle D and from G to semicircle E, intersecting at H. Prove that $HA \perp BC$.

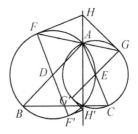

Figure 5.36

Analysis. Since a line and a circle have two intersection points, there are four possible pairs: $\{F, G\}$, $\{F', G'\}$, $\{F', G\}$, and $\{F, G'\}$. The last two pairs are not valid, and the first two pairs satisfy the conclusion. Therefore, we need to further determine the relationship between F and G. Obviously, $\triangle DAF \sim \triangle EAG$, and both triangles are isosceles. From $\angle DFA = \angle EAG$, it follows that points F, A, and G are collinear. Also, from $\angle DFA = \angle EGA$, we have $\angle HFG = \angle HGF$ and $(H - F)^2 = (H - G)^2$.

Proof.
$$(H - A)(B - C) + (F - A)(F - B) - (G - A)(G - C)$$
$$- 2\left(F - \frac{A+B}{2}\right)(F - H) + 2\left(G - \frac{A+C}{2}\right)(G - H)$$
$$+ [(H - F)^2 - (H - G)^2] = 0.$$

Example 12. In Figure 5.37, in $\triangle ABC$ with $AB = AC$, let D be the midpoint of side BC and E be a point outside $\triangle ABC$ such that $CE \perp AB$ and $BE = BD$. Let M be the midpoint of segment BE, and draw a line $MF \perp BE$, intersecting the circumcircle of $\triangle ABD$ at point F on the minor arc $\overparen{AD}$. Prove that $ED \perp DF$. (2010 China Girls' Mathematical Olympiad)

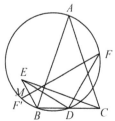

Figure 5.37

Note. In theory, any point on a plane can be represented using reference points A, B, C, and relevant parameters. However, calculating coordinates for points F and F' using constraint equations can be computationally intensive, exceeding what can be done manually. Additionally, there is no simple way to distinguish between F and F'. Therefore, this problem requires a combination of multiple methods to solve. First, prove $DE \perp DF$ and that F lies on the perpendicular bisector of BE. The rest is left to the reader. If you find a simpler proof based on point geometry, please let us know.

Exercise 5.3

1. In Figure 5.38, let AB be the diameter of the semicircle $\odot O$, and let P be a point on the diameter AB. Draw circle $\odot A$ with centre at point A and radius AP, which intersects the semicircle $\odot O$ at point C. Similarly, draw circle $\odot B$ with centre at point B and radius BP, which intersects

the semicircle $\odot O$ at point D. Let M be the midpoint of segment CD. Prove that MP is tangent to both circles $\odot A$ and $\odot B$. (2007 Math Weekly Cup National Secondary School Mathematics Competition)

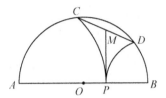

Figure 5.38

2. In Figure 5.39, construct equilateral triangles $\triangle BCD$, $\triangle CAE$, and $\triangle ABF$ externally to triangle $\triangle ABC$. Prove that $AD = EF \iff \angle A = 90°$. (2019 Danish Mathematical Olympiad Advanced Edition)

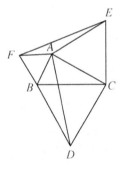

Figure 5.39

5.4 Trajectory Problems

Example 1. In Figure 5.40, consider $\triangle ABC$ with side AB of fixed length c. If the median of side BC has a fixed length of r, determine the locus of vertex C.

Solution 1. Take the midpoint O as the origin, with the line containing AB as the x-axis, establishing a

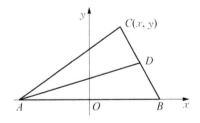

Figure 5.40

Cartesian coordinate system, as shown in Figure 5.40. Then, $A\left(-\frac{c}{2}, 0\right)$ and $B\left(\frac{c}{2}, 0\right)$. The coordinates of the midpoint D of BC are $\left(\frac{x+\frac{c}{2}}{2}, \frac{y}{2}\right)$. Since $|AD| = r$, we have

$$\left(\frac{x+\frac{c}{2}}{2} + \frac{c}{2}\right)^2 + \left(\frac{y}{2}\right)^2 = r^2,$$

which simplifies to $\left(x + \frac{3c}{2}\right)^2 + y^2 = 4r^2$. Therefore, the locus of C is a circle centred at $\left(-\frac{3c}{2}, 0\right)$ with a radius of $2r$.

Solution 2.

$$(C - (2A - B))^2 = 4\left(A - \frac{B+C}{2}\right)^2.$$

Example 2. In Figure 5.41, let AB be a chord of circle $x^2 + y^2 = a^2$. Given a fixed point $C(c, 0)$, such that $\angle ACB$ is a right angle, determine the locus of the midpoint P of AB ($|c| \neq |a|$).

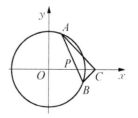

Figure 5.41

Solution. Let $O = 0$, then

$$4\left[\left(\frac{A+B}{2} - \frac{C}{2}\right)^2 - \frac{2A^2 - C^2}{4}\right]$$
$$+ (A^2 - B^2) - 2(C - A)(C - B) = 0.$$

When $2a^2 < c^2$, the locus does not exist. When $2a^2 = c^2$, the locus is a single point: $\left(\frac{c}{2}, 0\right)$. When $2a^2 > c^2$ and $a^2 \neq c^2$, the locus is a portion inside the known circle: $\left(\frac{A+B}{2} - \frac{C}{2}\right)^2 = \frac{2A^2 - C^2}{4}$.

Note. The equation can also be translated into the language of analytic geometry as $\left(x - \frac{c}{2}\right)^2 + y^2 = \frac{2a^2 - c^2}{4}$.

Example 3. A moving point P satisfies the condition that the sum of the squares of its distances to the vertices A and B of equilateral $\triangle ABC$ is equal to the square of its distance to vertex C. Find the locus of point P.

Solution. Let $C = 0$, then

$$[(P - (A+B))^2 - (A - B)^2] + [P^2 - (P - A)^2 - (P - B)^2]$$
$$+ (A^2 - B^2) - 2[A^2 - (A - B)^2] = 0.$$

The locus of point P is a circle centred at K with a radius equal to the side length AB, where K is symmetric to C about AB.

Example 4. In Figure 5.42, a line passing through the fixed point A intersects the fixed circle O at points M and N. Find the locus of the midpoint P of chord MN.

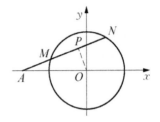

Figure 5.42

Solution. Let $O = 0$, then

$$\left[\left(\frac{M+N}{2} - \frac{A}{2}\right)^2 - \left(\frac{A}{2}\right)^2\right]$$

$$-\frac{M+N}{2}\left(\frac{M+N}{2} - A\right) = 0.$$

The desired locus is a part of the circle centred at the midpoint of AO, with a radius equal to half of AO, which lies inside the fixed circle O.

Exercise 5.4

1. A is a point inside $\odot O$, and B is a moving point on circle $\odot O$. A moving point P satisfies $PA = PB$, and $BO \perp BP$. Determine the locus of point P.
2. In $\triangle ABC$, side BC is fixed. Points M and N are the midpoints of sides AB and AC, respectively. If $BN = 2CM$, and BN intersects CM at point P, find the locus of point P as point A varies.

Appendix

Exercise Answers

A.1 Exercise 2.2

Problem 1. Given: $2E = C + D$, $2F = B + D$, $2G = A + B$, and $2H = E + G$.

Prove that there exists n such that $I = nF + (1-n)H$ and $2I = A + C$.

Proof. If $A + C = 2nF + 2(1-n)H$, then
$$A + C = n(B + D) + (1-n)\frac{A+B+C+D}{2}.$$
Therefore, $\frac{1-n}{2} = 1$, and $n + \frac{1-n}{2} = 0$, which implies that $n = -1$ makes the equation valid.

An alternative proof. It is evident that
$$\frac{C+D}{2} + \frac{A+B}{2} - \frac{B+D}{2} = \frac{A+C}{2}.$$
That is, $E + G - F = \frac{A+C}{2}$ and $2H - F = \frac{A+C}{2}$. Geometrically, this means that I is the midpoint of AC, and H is the midpoint of IF.

Problem 2. $\frac{|DC|}{|BD|} = \frac{|DC||BC|}{|BD||BC|} = \frac{|AC|^2}{|AB|^2} = k^2$, and $D = \frac{k^2B+C}{k^2+1}$.

$P = \frac{D+A}{2} = \frac{(1+k^2)A+k^2B+C}{2(1+k^2)}$, and $E = \frac{(1+k^2)A+C}{1+k^2+1}$, so $\frac{|AE|}{|EC|} = \frac{1}{1+k^2}$.

Problem 3. $B - A = 3(C - D)$.
$$F = kA + (1-k)D = kA + (1-k)\left(C - \frac{B-A}{3}\right)$$
$$= \frac{1+2k}{3}A + \frac{k-1}{3}B + (1-k)C,$$

and since F lies on median AC, then $\frac{1+2k}{3} = 1 - k$, which implies $k = \frac{2}{5}$ and $\frac{AF}{FD} = \frac{3}{2}$.

Problem 4. $L = \frac{aA+bB}{a+b}$, $P = \frac{aA+bB+aC}{a+b+a}$:

$$\frac{\frac{aA+bB+aC}{a+b+a} - C}{\frac{aA+bB}{a+b} - \frac{aA+bB+aC}{a+b+a}} - \frac{b}{a} = \frac{a+b}{a} - \frac{b}{a} = 1.$$

Problem 5. $\frac{aA+bB+cC}{a+b+c} = \frac{a+c}{a+b+c} \frac{aA+cC}{a+c} + \frac{b}{a+b+c} B$, so

$$\frac{|BI|}{|BE|} - \frac{|BD|}{|BA|} = \frac{a+c}{a+b+c} - \frac{\frac{a-b+c}{2}}{c} = \frac{b^2 + c^2 - a^2}{2c(a+b+c)}.$$

Problem 6. Let $B = 0$, $E = mA$, $D = A + C$, and

$$F = sE + (1-s)D = (1-s)C + (1-s+ms)A.$$

Solving $1 - s + ms = 0$ gives $s = \frac{1}{1-m}$, $F = \frac{m}{-1+m}C$. Let

$$M = tE + (1-t)C = mtA + (1-t)C,$$

$$M = rA + (1-r)F = \frac{m(1-r)}{-1+m}C + rA,$$

which gives $r = mt$, $t = \frac{1}{1-m+m^2}$, that is,

$$M = rA + (1-r)F = \frac{(A + C(-1+m))m}{1 - m + m^2},$$

$$(M - C)(E - C) - A^2 = \frac{C^2 + A^2(-1+m) - 2ACm}{1 - m + m^2}$$

$$= \frac{C^2(1 - 1 + m - m)}{1 - m + m^2} = 0.$$

Note. In the proof, several parameters were introduced. However, it is important to note that when point E is determined, i.e., when parameter m is fixed, the entire figure is determined. Therefore, other parameters need to be expressed in terms of m.

Problem 7. Let $O = xA + yB + zC$, where $x + y + z = 1$ and $0 < x, y, z < 1$. Consider

$$M = O + k(B - C) = xA + yB + zC + k(B - C)$$
$$= xA + (y+k)B + (z-k)C,$$

where $z - k = 0$, so $M = xA + (y + z)B$. Similarly,
$$N = xA + (y + z)C,$$
so $M - N = (y + z)(B - C)$. Likewise, the ratios for the other two line segments are $z + x$ and $x + y$, respectively. Adding these ratios, we have $y + z + z + x + x + y = 2$.

Problem 8. Let $\triangle ABC$ have the circumcentre as the origin. Assuming that the Euler line of $\triangle HBC$ intersects the line OH, then
$$m\frac{C + H + B}{3} + (1 - m)A = nH,$$
that is,
$$m\frac{A + 2B + 2C}{3} + (1 - m)A = n(A + B + C).$$
When $m = \frac{3}{4}$ and $n = \frac{1}{2}$,
$$\frac{3}{4}\frac{C + H + B}{3} + \frac{1}{4}A = \frac{A + B + C}{2}.$$
According to symmetry, it can be deduced that the Euler lines of $\triangle HCA$ and $\triangle HAB$ also pass through this intersection point.

Problem 9. Let $A = 0$ and $|BD| = |CE| = m$, then $D = \frac{c-m}{c}B$, $E = \frac{b-m}{b}C$,
$$G = \frac{mA + (c - m)B + (b - m)C}{m + (c - m) + (b - m)}, \quad F = \frac{B + C}{2},$$
$$M = 2F - G = \frac{bB + cC - mA}{b + c - m},$$
so M lies on the bisector of $\angle BAC$.

A.2 Exercise 2.3

Problem 1. $2(F - C) = B - A$, and
$$I = kB + (1 - k)F = kB + (1 - k)\left(C + \frac{B - A}{2}\right)$$
$$= \frac{k - 1}{2}A + \frac{1 + k}{2}B + (1 - k)C.$$
Since I lies on AE, we have $\frac{\frac{1+k}{2}}{1-k} = \frac{2}{1}$, which gives $k = \frac{3}{5}$, and $I = \frac{-A + 4B + 2C}{5}$. Assuming $B = 0$, we get $(I - A)(I - C) = \frac{6}{25}(A^2 - C^2) = 0$.

Problem 2. Let $d = |BD|$, $D = \frac{d}{a}C + \frac{a-d}{a}B$,
$$E = \frac{c-(a-d)}{c}B + \frac{a-d}{c}A, \quad F = \frac{d}{b}A + \frac{b-d}{b}C,$$
$$\left(D - \frac{E+F}{2}\right)(E - F) = \frac{(a-b-c)(a+b+c)(a-d)d(b-c)}{2abc}.$$
Therefore, $DM \perp EF \iff AB = AC$.

Problem 3. Let $A = 0$, $P = tC$, $U = \frac{bB+cC}{b+c}$,
$$(U - P)O = \left(\frac{bB + cC}{b+c} - tC\right)O = \frac{bBO + cCO}{b+c} - tCO$$
$$= \frac{\frac{bc^2}{2} + \frac{cb^2}{2}}{b+c} - \frac{tb^2}{2} = \frac{1}{2}b(c - bt) = 0;$$
$UP \perp AO \iff t = \frac{c}{b} \iff AB = AP$.

Problem 4. Let $A = 0, |B| = |C| = \sqrt{2}|D|, \cos\left(\frac{\pi}{4} - \frac{\pi}{6}\right) = \frac{1+\sqrt{3}}{2\sqrt{2}}$. We have $\left(D - \frac{B}{2}\right)B = 0$,
$$\left(D - \frac{B}{2}\right)(D - C) = \frac{BC}{2} - \frac{BD}{2} - CD + D^2$$
$$= \frac{\sqrt{3}}{2}\frac{2D^2}{2} - \frac{D^2}{2} - \sqrt{2}\frac{1+\sqrt{3}}{2\sqrt{2}}D^2 + D^2 = 0,$$
so $DC \parallel AB$.

Problem 5. Let $A = 0, D = k\left(\frac{bB+cC}{b+c}\right)$,
$$E = C + m(B - D) = C + m\left[B - k\left(\frac{bB + cC}{b+c}\right)\right].$$
Because E lies on AB, we have $1 - mk\frac{c}{b+c} = 0$, $m = \frac{b+c}{ck}$,
$$E = \frac{b+c}{ck}\left(1 - \frac{kb}{b+c}\right)B, \quad G = \frac{E+C}{2} = \frac{bB + cB - bkB + ckC}{2ck}.$$
By symmetry,
$$H = \frac{bC + cC - ckC + bkB}{2bk},$$
$$D(G - H) = \frac{(bB - cC)(bB + cC)(1-k)}{2bc} = \frac{(b^2c^2 - b^2c^2)(1-k)}{2bc} = 0.$$

Problem 6. Let $T = \frac{bB+cC}{b+c}$, $M = \frac{B+C}{2}$. According to the ratio relationship,
$$\frac{D-B}{A-B} = \frac{M-B}{T-B},$$
$$D = \frac{\frac{B+C}{2} - B}{\frac{bB+cC}{b+c} - B}(A-B) + B = \frac{(b+c)A + (c-b)B}{2c},$$
$$|BD| = \frac{(b+c)c}{2c} = \frac{b+c}{2}.$$
Similarly, $|CE| = \frac{b+c}{2}$.

Note. Alternatively, D can be found as follows:
$$T = \frac{bB + cC}{b+c} = \frac{bB + c(2M-B)}{b+c},$$
$$M = \frac{(b+c)T - (b-c)B}{2c},$$
$$D = \frac{(b+c)A - (b-c)B}{2c}.$$

Problem 7. Let $D = tB + (1-t)C$, $E = sB + (1-s)C$. Then, by the ratio relationship $\frac{C-F}{C-A} = \frac{C-E}{C-D}$, we have
$$F = C - \frac{C-E}{C-D}(C-A) = \frac{sA - sC + tC}{t},$$
and by the ratio relationship $\frac{B-G}{B-A} = \frac{B-E}{B-D}$,
$$G = B - \frac{B-E}{B-D}(B-A) = \frac{(s-1)A + (t-s)B}{t-1},$$
$$F - E + G - E = \frac{(A + (t-1)C - tB)(s(2t-1) - t)}{(t-1)t},$$
$$2(A-D) = 2(A + (t-1)C - tB),$$
$$|EF| + |EG| - 2|AD| = \frac{(s-t)(2t-1)(A + (t-1)C - tB)}{(t-1)t}.$$

Since D and E do not coincide, A, B, C are not collinear, and so only $2t - 1 = 0$, and D is the midpoint of BC.

Note. Alternatively, we can use the ratio relationships $sD - tE = (s-t)C = sA - tF$ and $(1-s)D - (1-t)E = B(t-s) = (1-s)A - (1-t)G$ to solve for F and G.

Problem 8. Let $C = 0$:

$$\left(\frac{aA+bB+cC}{a+b+c} - \frac{A+B}{2}\right)\left(\frac{aA+bB+cC}{a+b+c} - B\right)$$
$$= \frac{(aA - aB - cB)(aA - bA - cA - aB + bB - cB)}{2(a+b+c)^2}.$$

The numerator is $a(a-b-c)A^2 + (a+c)(a-b+c)B^2$

$$= a(a-b-c)b^2 + (a+c)(a-b+c)a^2$$
$$= -a(-a^3 + a^2b - ab^2 + b^3 - 2a^2c + abc + b^2c - ac^2)$$
$$= -a[-a(c^2 - b^2) + a^2b - ab^2 + b(c^2 - a^2)$$
$$\quad - 2a^2c + abc + (c^2 - a^2)c - ac^2]$$
$$= ac(3a - b - c)(a + c) = 0.$$

Thus, $c = 3a - b$, and $a^2 + b^2 - (3a-b)^2 = -2a(4a-3b) = 0$, and it follows that

$$BC : CA : AB = 3 : 4 : 5.$$

Problem 9. Let $A = 0$:

$$(D - E)(D - F)$$
$$= \left(\frac{bB+cC}{b+c} - \frac{cC+aA}{c+a}\right)\left(\frac{bB+cC}{b+c} - \frac{bB+aA}{b+a}\right)$$
$$= \frac{b^2(a-c)(a+c)B^2 + 2bc(a^2+bc)BC + (a-b)(a+b)c^2C^2}{(a+b)(a+c)(b+c)^2}.$$

Simplifying the numerator gives

$$b^2(a-c)(a+c)c^2 + bc(a^2+bc)(b^2+c^2-a^2) + (a-b)(a+b)c^2b^2$$
$$= -a^2bc(a^2 - b^2 - bc - c^2) = 0.$$

Therefore, $DE \perp DF \iff \angle A = 120°$.

Problem 10. Let $A = 0$, $|AB| = 1$, $P = tB + (1-t)C$, $M = sB$:

$$M^2 - (M-P)^2 = -2(s-t)(t-1)BC - (t-1)^2 C^2 + (2s-t)tB^2$$
$$= -(s-t)(t-1) - (t-1)^2 + (2s-t)t = 0.$$

It gives $s = \frac{1-t+t^2}{1+t}$, $|BM| = 1 - \frac{1-t+t^2}{1+t} = \frac{t(2-t)}{1+t}$. Similarly, $|CN| = \frac{(1-t)(2-(1-t))}{1+(1-t)}$. Also,

$$|BP| \cdot |PC| = t(1-t),$$

$$|BM||CN| = \frac{t(2-t)}{1+t} \frac{(1-t)(2-(1-t))}{1+(1-t)} = t(1-t).$$

Problem 11.

$$E = \frac{\left(\frac{a+b+c}{2} - c\right)A + \left(\frac{a+b+c}{2} - a\right)C}{\left(\frac{a+b+c}{2} - c\right) + \left(\frac{a+b+c}{2} - a\right)} = \frac{A(a+b-c) + (-a+b+c)C}{2b},$$

$$D = \frac{A+C}{2}, \quad F = \frac{aA - bB + cC}{a-b+c}, \quad E - B = \frac{a-b+c}{b}(F-D).$$

Therefore, $BE \parallel DF$.

Problem 12. Proof 1: Let $O = 0$. Then, in the expression

$$\frac{-aA + bB + cC}{-a+b+c} \left(\frac{aA+cC}{a+c} - \frac{aA+bB}{a+b} \right), \text{ the numerator is}$$

$$a^2(b-c)A^2 + b^2(a+c)B^2 - c^2(a+b)C^2 - bc(b-c)BC$$
$$- ab(a+b)AB + ac(a+c)AC = a^2(b-c)R^2 + b^2(a+c)R^2 - c^2(a+b)R^2$$
$$- bc(b-c)\frac{2R^2 - a^2}{2}$$
$$- ab(a+b)\frac{2R^2 - c^2}{2} + ac(a+c)\frac{2R^2 - b^2}{2}$$
$$= 0.$$

Proof 2: Let $A = 0$:

$$\left(0 - \frac{-aA + bB + cC}{-a+b+c}\right)\left(\frac{aA + cC}{a+c} - \frac{aA + bB}{a+b}\right)$$

$$= \frac{cb^2}{2}\frac{1}{a+c} - \frac{bc^2}{2}\frac{1}{a+b}$$

$$- \frac{bcBC + c^2C^2}{-a+b+c}\frac{1}{a+c} + \frac{b^2B^2 + bcBC}{-a+b+c}\frac{1}{a+b}$$

$$= \frac{cb^2}{2}\frac{1}{a+c} - \frac{bc^2}{2}\frac{1}{a+b}$$

$$- \frac{bc\frac{b^2+c^2-a^2}{2} + c^2b^2}{-a+b+c}\frac{1}{a+c}$$

$$+ \frac{b^2c^2 + bc\frac{b^2+c^2-a^2}{2}}{-a+b+c}\frac{1}{a+b}$$

$$= 0.$$

Problem 13.

$$P = tB + (1-t)C, \quad F = tB + (1-t)A, \quad E = tA + (1-t)C,$$

$$D = \frac{B+C}{2}, \quad O = \frac{t(1-t)A + t^2B + (1-t)^2C}{t(1-t) + t^2 + (1-t)^2}.$$

Let $K = mO + (1-m)P$

$$= \frac{\begin{array}{c}-m(1-t)tA - t(-1+m+t-2mt-t^2+mt^2)B \\ +(-1+t)(-1+t-t^2+mt^2)C\end{array}}{1-t+t^2}$$

be on the median AD, the coefficients of B and C be equal, and

$$-t(-1+m+t-2mt-t^2+mt^2) = (-1+t)(-1+t-t^2+mt^2).$$

Then, it gives $m = \frac{t^2-t+1}{(t-1)t}$, and verification shows that $mO + (1-m)P = B + C - A$.

Problem 14. Let $A = 0$, $D = \frac{bB+cC}{b+c}$, $E = \frac{bA+cC}{b+c}$, $F = \frac{bB+cA}{b+c}$, $G = \frac{bcA+b^2B+c^2C}{bc+b^2+c^2}$. Let

$$(A-G)(B-C) - k(A-B)(A-C)$$

$$= \frac{-b^2B^2 + c^2C^2 + BC(b^2 - c^2 - b^2k - bck - c^2k)}{b^2 + bc + c^2} = 0.$$

When $k = \frac{b^2-c^2}{b^2+bc+c^2}$,

$$(A-G)(B-C) - \frac{b^2-c^2}{b^2+bc+c^2}(A-B)(A-C) = \frac{-b^2B^2+c^2C^2}{b^2+bc+c^2} = 0.$$

Problem 15. Let $A = 0, P = (1-2t)A + tB + tC, E = \frac{(1-2t)A+tC}{(1-2t)+t}, F = \frac{(1-2t)A+tB}{(1-2t)+t}$. Let $(B-E)^2 - (C-F)^2 - k(B^2 - C^2) = \frac{(B^2-C^2)(1-k-2t+2kt-kt^2)}{(-1+t)^2}$. Solving $1 - k - 2t + 2kt - kt^2 = 0$ gives $k = \frac{1-2t}{(1-t)^2}$:

$$(B-E)^2 - (C-F)^2 - \frac{1-2t}{(1-t)^2}(B^2 - C^2) = 0.$$

Problem 16. Let $A = 0$, $D = dB$, $E = eC$, $M = \frac{B+E}{2}$, $N = \frac{C+D}{2}$:

$$tM + (1-t)N = B\left(\frac{d(1-t)}{2} + \frac{t}{2}\right) + C\left(\frac{1-t}{2} + \frac{et}{2}\right).$$

When $\frac{d(1-t)}{2} + \frac{t}{2} = 0$, $t = \frac{d}{-1+d}$, then $Q = \frac{1-de}{2-2d}C$; when $\frac{1-t}{2} + \frac{et}{2} = 0$, $t = \frac{1}{1-e}$, then $P = \frac{1-de}{2-2e}B$:

$$P^2 - Q^2 = \frac{(1-de)^2}{4(1-d)^2(1-e)^2}[(1-d)^2B^2 - (1-e)^2C^2],$$

$$(B-D)^2 - (C-E)^2 = (1-d)^2B^2 - (1-e)^2C^2.$$

Since $1 - de \neq 0$, we have $BD = CE \iff AP = AQ$.

Problem 17. Let $A = 0$, $\quad E = tA + (1-t)B$, $\quad F = sA + (1-s)C$,

$$P = \frac{E+F}{2} = \frac{C(1-s) + B(1-t) + At + As}{2},$$

$$D = \frac{C(1-s) + B(1-t)}{(1-s) + (1-t)},$$

$(D-E)^2 - (B-E)^2 - (D-F)^2 + (C-F)^2 = \frac{(B^2-C^2)(s+t-2st)}{-2+s+t}$. Note that $DE^2 - BE^2 = DF^2 - CF^2$ does not necessarily imply $AB = AC$; it could also be $s + t - 2st = 0$.

Problem 18. Let $B = 0$, $M = \frac{A+C}{2}, L = \frac{aA+cC}{a+c}$,

$$(a+c)L - 2cM = (a-c)A, \quad M = \frac{(a+c)L - (a-c)A}{2c},$$

$$D = \frac{(a+c)L - (a-c)B}{2c} = \frac{aA - (a-c)B + cC}{2c},$$

$$(a+c)L - 2aM = (-a+c)C,$$

$$L = \frac{2aM + (-a+c)C}{a+c},$$

$$E = \frac{2aM + (-a+c)B}{a+c} = \frac{cB + a(A - B + C)}{a+c},$$

$$(E - D)(B - L) = (a^3 - a^2c)A^2 + (-ac^2 + c^3)C^2$$

$$= c^2(a^3 - a^2c) + (-ac^2 + c^3)a^2 = 0.$$

Problem 19. Let $B = 0$, then $\frac{1}{2}\frac{b+c}{b}\left(A - \frac{bB+cC}{b+c}\right)^2 - c^2$

$$= \frac{(b+c)\left(A^2 + \frac{c^2 C^2}{(b+c)^2} - 2\frac{c}{b+c}AC\right)}{2b} - c^2$$

$$= \frac{(b+c)\left(c^2 + \frac{c^2 a^2}{(b+c)^2} - \frac{c}{b+c}(a^2 + c^2 - b^2)\right)}{2b} - c^2$$

$$= -\frac{c(a^2 - b^2 + c^2)}{2(b+c)}.$$

So, $\frac{AB^2}{AD^2} = \frac{1}{2}\frac{BC}{DC} \iff BA \perp BC$.

Problem 20. Let $A = 0$, then

$$\left(C - \frac{aA + bB + cC}{a+b+c}\right)^2 = \left(\frac{(a+b)C - bB}{a+b+c}\right)^2$$

$$= \frac{-b(a+b)(-a^2 + b^2 + c^2) + b^2(a+b)^2 + b^2c^2}{(a+b+c)^2}$$

$$= \frac{ab(a+b-c)}{a+b+c}.$$

Similarly,

$$\left(A - \frac{aA + bB + cC}{a+b+c}\right)^2 = \frac{2b^2c^2 + bc(b^2 + c^2 - a^2)}{(a+b+c)^2} = \frac{bc(-a+b+c)}{a+b+c}.$$

So, $\frac{CI^2}{AI^2} = \frac{BC}{AB}\frac{MC}{AM}$.

Problem 21. Let $A = 0$, then $|BD| = |CE| = m$, $D = \frac{c-m}{c}B$, $L = \frac{c-b}{c}B$,

$$E = \frac{b-m}{b}C, \quad F = \frac{mA + (c-m)B + (b-m)C}{m + (c-m) + (b-m)}, \quad T = \frac{bB + cC}{b+c},$$

$$L - F = \frac{(m-b)(bB+cC)}{c(b+c-m)}.$$

So, $AT \parallel LF$.

Problem 22. Let $O = 0$, then

$OI^2 - HI^2$

$$= \left(\frac{aA + bB + cC}{a+b+c}\right)^2 - \left(\frac{aA + bB + cC}{a+b+c} - A - B - C\right)^2$$

$$= \frac{(a-b-c)A^2 - (a-b+c)B^2 - (a+b-c)C^2 - 2ABc - 2ACb - 2BCa}{a+b+c}.$$

Substituting $A^2 = B^2 = C^2 = R^2$, $2AB = 2R^2 - c^2$, $2BC = 2R^2 - a^2$, and $2CA = 2R^2 - b^2$ into the numerator, we get

$$a^3 + b^3 + c^3 - 3aR^2 - 3bR^2 - 3cR^2 = 0.$$

Substituting $2R\sin 60° = a$, i.e., $3R^2 = a^2$, we get

$$(b+c)(b^2 + c^2 - a^2 - bc) = (b+c)(b^2 + c^2 - a^2 - 2bc\cos 60°) = 0.$$

Problem 23. $ID = IE$

$$\iff \left(\frac{aA + bB + cC}{a+b+c} - \frac{aA + cC}{a+c}\right)^2 - \left(\frac{aA + bB + cC}{a+b+c} - \frac{aA + bB}{a+b}\right)^2$$

$$= \frac{\frac{b^2(a(A-B)+c(-B+C))^2}{(a+c)^2} - \frac{c^2(a(A-C)+b(B-C))^2}{(a+b)^2}}{(a+b+c)^2}$$

$$= \frac{\frac{b^2ac(a^2-b^2+2ac+c^2)}{(a+c)^2} - \frac{c^2ab(a^2+2ab+b^2-c^2)}{(a+b)^2}}{(a+b+c)^2}$$

$$= \frac{abc(b-c)(a^2 - b^2 + bc - c^2)}{(a+b)^2(a+c)^2} = 0,$$

where $b = c \iff \angle B = \angle C$, and $a^2 - b^2 + bc - c^2 = 0 \iff \angle A = 60°$. Moreover,

$$(a(A - B) + c(-B + C))^2$$
$$= a^2(A - B)^2 + c^2(-B + C)^2 + 2ac(B - A)(B - C)$$
$$= a^2c^2 + c^2a^2 + ac(a^2 + c^2 - b^2)$$
$$= ac(a^2 - b^2 + 2ac + c^2),$$

and similarly,

$$(a(A - C) + b(B - C))^2 = a^2(A - C)^2 + b^2(B - C)^2 + 2ab(A - C)(B - C)$$
$$= a^2b^2 + b^2a^2 + ab(a^2 + b^2 - c^2)$$
$$= ab(a^2 + 2ab + b^2 - c^2).$$

Problem 24. Let the midpoint K of AB be the origin, $|KB| = m$, $|KP| = n$, $O = -\frac{m^2}{n}\frac{P}{n}$, $A = -B$, $D = kB$, $C = tP + (1-t)A$:

$$(O - D)(B - C) = -2kB^2 + ktB^2 + \frac{m^2t}{n^2}P^2 = -2km^2 + ktm^2 + m^2t.$$

Solving the equation $-2km^2 + ktm^2 + m^2t = 0$, we have $t = \frac{2k}{1+k}$.

$$rP + (1-r)D - \frac{B+C}{2} = \left(k - kr - \frac{t}{2}\right)B + \left(r - \frac{t}{2}\right)P = 0.$$

Solving the equation $k - kr - \frac{t}{2} = r - \frac{t}{2} = 0$, we have $t = \frac{2k}{1+k}$. Therefore, the proposition is proved.

Problem 25. Let $D = aA + bB + (1 - a - b)C$,

$$E = \frac{D - aA}{1 - a} = \frac{bB + (1 - a - b)C}{1 - a},$$

$$I = \frac{D - bB}{1 - b} = \frac{aA + (1 - a - b)C}{1 - b}.$$

When $AB \parallel CD$ (i.e., $a = -b$),

$$\left(\frac{aA + (1 - a - b)C}{1 - b} - C\right)^2 - \left(\frac{aA + (1 - a - b)C}{1 - b} - A\right)(C - A)$$
$$= \frac{a^2b^2}{(-1+b)^2} - \frac{b^2(-1 + a + b)}{-1 + b} = \frac{b^2(-1 + a + a^2 + 2b - ab - b^2)}{(-1+b)^2} = 0,$$

we get

$$-1 - a + a^2 = 0.$$

To make $E + D + C = I + A + B$, i.e.,

$$\frac{bB + (1 - a - b)C}{1 - a} + aA + bB + (1 - a - b)C + C$$

$$= \frac{aA + (1 - a - b)C}{1 - b} + A + B.$$

Substituting $b = -a$, $-1 - a + a^2 = 0$ into the equation gives

$$\frac{(-1 - a + a^2)(-A + aA - B - aB + 2C)}{(-1 + a)(1 + a)} = 0.$$

Problem 26. Let $D = \frac{xA+yB+zC}{x+y+z}$, $M = \frac{A+B}{2}$, $N = \frac{C+D}{2}$, $L = \frac{xA+zC}{x+z}$,

$$K = B + t(A - D) = \frac{t(y + z)A + (x + y - ty + z)B - tzC}{x + y + z}.$$

Since $AK \parallel CB$, thus $x + y - ty + z = tz$, which gives $t = \frac{x+y+z}{y+z}$, $K = \frac{(y+z)A+zB-zC}{y+z}$,

$$M - N = \frac{(y + z)A + (x + z)B + (-x - y - 2z)C}{2(x + y + z)},$$

$$K - L = \frac{z((y + z)A + (x + z)B + (-x - y - 2z)C)}{(x + z)(y + z)},$$

so $LK \parallel NM$.

Problem 27. Let $K = \frac{xA+yB+zC}{x+y+z}$, $D = \frac{yB+zC}{y+z}$, $E = \frac{xA+zC}{x+z}$, $F = \frac{xA+yB}{x+y}$,

$$Q = \frac{xA + 2yB + zC}{x + 2y + z} = \frac{2y}{x + 2y + z}B + \left(1 - \frac{2y}{x + 2y + z}\right)E$$

$$= \frac{y + z}{x + 2y + z}D + \left(1 - \frac{y + z}{x + 2y + z}\right)F.$$

Similarly, $R = \frac{xA+yB+2zC}{x+y+2z}$,

$$X = \frac{2xA + 3yB + 3zC}{2x + 3y + 3z} = \frac{2x}{2x + 3(y + z)}A + \left(1 - \frac{2x}{2x + 3(y + z)}\right)D$$

$$= \frac{x + 2y + z}{2x + 3(y + z)}Q + \left(1 - \frac{x + 2y + z}{2x + 3(y + z)}\right)R,$$

$$\frac{AD}{DX} + \frac{1}{2} = \frac{3(x + y + z)}{2x}.$$

Similarly, the other two expressions can be obtained. We have $\frac{1}{\frac{AD}{DX}+\frac{1}{2}} + \frac{1}{\frac{BE}{EY}+\frac{1}{2}} + \frac{1}{\frac{CF}{FZ}+\frac{1}{2}} = \frac{2}{3}$.

Problem 28. Let $O = \frac{xA+yB+zC}{x+y+z}$, $D = \frac{yB+zC}{y+z}$, $E = \frac{xA+zC}{x+z}$, $F = \frac{xA+yB}{x+y}$:

$$P = \frac{x+z}{2x+y+z}E + \left(1 - \frac{x+z}{2x+y+z}\right)F$$

$$= \frac{x}{2x+y+z}A + \left(1 - \frac{x}{2x+y+z}\right)O,$$

$$\frac{OP}{AP}\left(\frac{AF}{BF} + \frac{AE}{CE}\right) = \frac{\frac{x}{2x+y+z}}{1 - \frac{x}{2x+y+z}}\left(\frac{y}{x} + \frac{z}{x}\right) = \frac{y+z}{x+y+z},$$

$$\frac{y+z}{x+y+z} + \frac{z+x}{x+y+z} + \frac{x+y}{x+y+z} = 2.$$

Problem 29. Let $D = \frac{xA+yB+zC}{x+y+z}$, $J = \frac{A+C}{2}$, $K = \frac{B+D}{2}$, $E = \frac{xA+yB}{x+y}$, $F = \frac{yB+zC}{y+z}$, $G = \frac{E+D}{2}$, $H = \frac{B+F}{2}$. Calculating, it yields

$$I = \frac{x(2x+2y+z)A - z(x+2y+z)B + z(x+y)C}{(2x-z)(x+y+z)}$$

$$= \frac{2y+z}{z-2x}B + \left(1 - \frac{2y+z}{z-2x}\right)G$$

$$= \frac{2x+2y+z}{2x-z}D + \left(1 - \frac{2x+2y+z}{2x-z}\right)H,$$

$$K - J = \frac{-(y+z)A + (x+2y+z)B - (x+y)C}{2(x+y+z)}.$$

Let $A = 0$, then evidently $AI \parallel JK$.

Problem 30. $F = \frac{xA+yB+zC}{x+y+z}$, $D = \frac{yB+zC}{y+z}$, $E = \frac{xA+yB}{x+y}$,. Let $G = tD + (1-t)E = \frac{(1-t)(xA+yB)}{x+y} + \frac{t(yB+zC)}{y+z}$. Solving the equation $\frac{\frac{(1-t)x}{x+y}}{\frac{tz}{y+z}} = \frac{x}{z}$ yields $t = \frac{y+z}{x+2y+z}$. So, $G = \frac{xA+2yB+zC}{x+2y+z}$,

$$M = \frac{y+z}{x+2y+z}B + \left(1 - \frac{y+z}{x+2y+z}\right)E = \frac{xA+(2y+z)B}{x+2y+z},$$

$$H = \frac{y+z}{x+2y+z}C + \left(1 - \frac{y+z}{x+2y+z}\right)E = \frac{xA+yB+(y+z)C}{x+2y+z},$$

$N = \frac{xA+(2y+z)C}{x+2y+z}$. So, $G - H - (H - N) = 0$.

Problem 31. Let $D = \frac{\frac{a+b-c}{2}B + \frac{a-b+c}{2}C}{a}$, $I = \frac{aA+bB+cC}{a+b+c}$, and

$$E = 2I - D$$

$$= \frac{4a^2 A + (-a+b+c)(a-b+c)B + (-a+b+c)(a+b-c)C}{2a(a+b+c)}.$$

This implies that

$$\frac{|BF|}{|FC|} = \frac{(-a+b+c)(a+b-c)}{(-a+b+c)(a-b+c)} = \frac{a+b-c}{a-b+c}, \quad \frac{|BF|}{|BC|} = \frac{a+b-c}{2a},$$

and

$$\frac{|CD|}{|BC|} = \frac{a+b-c}{2a},$$

so $BF = DC$.

Problem 32. Let $B = 0$, and the inscribed circle touches BC at D. Let $I = \frac{aA+bB+cC}{a+b+c}$, $M = \frac{B+C}{2}$, and let $E = tI + (1-t)M$:

$$(A-E)(B-C) = \frac{2AC(-a-b-c+at) + C^2(a+b+c-at-bt+ct)}{2(a+b+c)}.$$

Solving the equation $(a^2 + c^2 - b^2)(-a-b-c+at) + a^2(a+b+c-at-bt+ct) = 0$ gives $t = \frac{b+c}{a}$,

$$E = tI + (1-t)M = \frac{2a(b+c)A + (a^2+b^2-c^2)B + (a^2-b^2+c^2)C}{2a(a+b+c)},$$

and

$$I - D = \frac{aA+bB+cC}{a+b+c} - \frac{\frac{a+b-c}{2}B + \frac{a-b+c}{2}C}{a}$$

$$= \frac{a^2(2A-B-C) - (b^2-c^2)(B-C)}{2a(a+b+c)} = A - E.$$

Note. Consider using the identical method. Consider the parallelogram $AIDE_1$, where $E_1 = A + D - I$. Verify that M, I, and E_1 are collinear.

The essence of this problem is the following identity:
$$\frac{2a(b+c)A + (a^2+b^2-c^2)B + (a^2-b^2+c^2)C}{2a(a+b+c)}$$
$$= A + \frac{\frac{a+b-c}{2}B + \frac{a-b+c}{2}C}{a} - \frac{aA+bB+cC}{a+b+c}$$
$$= \frac{b+c}{a}\frac{aA+bB+cC}{a+b+c} + \left(1 - \frac{b+c}{a}\right)\frac{B+C}{2}.$$

Problem 33. Let $B = 0$,
$$M = \frac{\frac{a+b-c}{2}B + \frac{a-b+c}{2}C}{a}, \quad N = \frac{\frac{a-b+c}{2}A + \frac{-a+b+c}{2}B}{c},$$
$$I = \frac{aA+bB+cC}{a+b+c}.$$

Let $P = kM + (1-k)N$
$$= \frac{-a(a-b+c)(-1+k)A + (-a^2+ab+ac+a^2k-abk+bck-c^2k)B + c(a-b+c)kC}{2ac}.$$

Solve the equation
$$\frac{-a(a-b+c)(-1+k)}{-a^2+ab+ac+a^2k-abk+bck-c^2k} = \frac{a}{b}$$
to find $k = \frac{a-b}{a-b+c}$:
$$P = kM + (1-k)N = \frac{b(B-C) + a(A+C)}{2a},$$
$$(P-C)(P-A) = \frac{-a^2A^2 + 2a^2AC + (b^2-a^2)C^2}{4a^2}$$
$$= \frac{-a^2c^2 + a^2(a^2+c^2-b^2) + (b^2-a^2)a^2}{4a^2} = 0.$$

An alternative proof. Let symmetric point A_1 of A about CI be $\frac{(a-b)C+bB}{a}$. CI intersects AA_1 at $P_1 = \frac{aA+bB+(a-b)C}{2a}$. It suffices to prove that P_1 lies on MN.

Problem 34. Let $P = \frac{xA+yB+zC}{x+y+z}$, $D = \frac{yB+zC}{y+z}$, $E = \frac{xA+zC}{x+z}$, $F = \frac{xA+yB}{x+y}$,
$$X = \frac{E+F}{2} = \frac{\frac{xA+zC}{x+z} + \frac{xA+yB}{x+y}}{2},$$

Appendix: Exercise Answers   187

where the coefficients of B and C are in the ratio $\frac{y}{x+y} : \frac{z}{x+z}$;

$$Y = \frac{D+F}{2} = \frac{\frac{yB+zC}{y+z} + \frac{xA+yB}{x+y}}{2},$$

where the coefficients of A and C are in the ratio $\frac{x}{x+y} : \frac{z}{y+z}$. Therefore, Q in terms of the ratios with respect to A, B, and C is

$$\frac{x(y+z)}{x+y} : \frac{y(x+z)}{x+y} : z,$$

$$Q = \frac{\frac{x(y+z)}{x+y}A + \frac{y(x+z)}{x+y}B + zC}{\frac{x(y+z)}{x+y} + \frac{y(x+z)}{x+y} + z}$$

$$= \frac{x(y+z)A + y(x+z)B + z(x+y)C}{2(xy+yz+zx)}.$$

By symmetry, Q is also on CZ.

Problem 35. Let $B = 0$, $|BD| = d$, $|BE| = e$, $C = \frac{d+e}{e}E$, $A = \frac{2d+e}{d}D$. Let

$$G = mA + (1-m)E = nC + (1-n)D,$$

then

$$mA + (1-m)E - nC - (1-n)D$$

$$= \left(-1 + \left(2 + \frac{e}{d}\right)m + n\right)D - \frac{dn + e(-1+m+n)}{e}E.$$

Solving the system of equations $-1 + \left(2 + \frac{e}{d}\right)m + n = 0$ and $dn + e(-1 + m + n) = 0$ gives

$$m = \frac{d^2}{2d^2 + 2de + e^2},$$

$$G = mA + (1-m)E = \frac{(2d^2 + de)D + (d^2 + 2de + e^2)E}{2d^2 + 2de + e^2},$$

$$(F - G)(C - D) = (D + E - G)(C - D)$$
$$= -de^2D^2 - e^3D^2 + 2de^2DE + e^3DE + d^3E^2 + d^2eE^2$$
$$= -d^3e^2 - d^2e^3 + 0 + 0 + d^3e^2 + d^2e^3 = 0.$$

Problem 36. Analysis: when M coincides with A, I also coincides with A, and D becomes the midpoint of AC. Thus, $\angle BDI = 90°$ and $\angle IBD = 30°$, providing the direction for progression.

Let $AB = 1$, $A = 0$, $M = kB$, $I = k\frac{A+B+C}{3}$, $D = \frac{C+M}{2}$:

$$(D-B)(D-I) = \frac{1}{12}(-6BC + 3C^2 - 2kB^2 + 8kBC$$
$$- 2kC^2 + k^2B^2 - 2k^2BC)$$
$$= \frac{1}{12}(-3 + 3 - 2k + 4k - 2k + k^2 - k^2) = 0,$$

$$4(I-D)^2 - (I-B)^2 = \frac{1}{9}(-9B^2 + 9C^2) + \frac{1}{9}(6B^2 + 12BC - 12C^2)k$$
$$+ \frac{1}{9}(-6BC + 3C^2)k^2 = 0.$$

So, $\angle BDI = 90°$, $\angle IBD = 30°$, and $\angle DIB = 60°$.

Problem 37. Let $A = 0$, $U = \frac{bB+cC}{b+c}$, $I = tO$:

$$(U-I)(B-C) = \left(\frac{bB^2 + cBC}{b+c} - tBO\right) - \left(\frac{bBC + cC^2}{b+c} - tCO\right)$$

$$= \left(\frac{bc^2 + c\frac{b^2+c^2-a^2}{2}}{b+c} - \frac{c^2}{2}t\right) - \left(\frac{b\frac{b^2+c^2-a^2}{2} + cb^2}{b+c} - \frac{b^2}{2}t\right)$$

$$= \frac{(b-c)(a^2 - b^2 - 2bc - c^2 + b^2t + 2bct + c^2t)}{2(b+c)}.$$

$IA = IU \iff \frac{bB+cC}{b+c}\left(tO - \frac{\frac{bB+cC}{b+c}}{2}\right) = 0$. The numerator of this expression is

$$(bB + cC)(2(b+c)tO - (bB + cC))$$
$$= 2bt(b+c)BO - (b^2B^2 + bcBC) + (2ct(b+c)CO - (bcBC + c^2C^2))$$
$$= bc^2(b+c)t - \left(b^2c^2 + bc\frac{b^2+c^2-a^2}{2}\right)$$
$$+ \left(b^2c(b+c)t - \left(bc\frac{b^2+c^2-a^2}{2} + b^2c^2\right)\right)$$
$$= bc(a^2 - b^2 - 2bc - c^2 + b^2t + 2bct + c^2t).$$

Therefore, $IU \perp BC \iff a^2 - b^2 - 2bc - c^2 + b^2t + 2bct + c^2t = 0 \iff IA = IU$.

Problem 38. Let $A = 0$, $E = \frac{aA+(c-a)B}{c}$, $F = \frac{aA+(b-a)C}{b}$, $K = \frac{aA+(c-a)B+(b-a)C}{a+(c-a)+(b-a)}$:

$$(O-K)(B-C) = \frac{aAB - aB^2 + cB^2 - aAC + bBC - cBC + aC^2 - bC^2}{a - (b+c)}$$

$$+ BO - CO$$

$$= \frac{0 - ac^2 + cc^2 - 0 + b\frac{b^2+c^2-a^2}{2} - c\frac{b^2+c^2-a^2}{2} + ab^2 - bb^2}{a - (b+c)}$$

$$+ c\frac{c-a}{2} - b\frac{b-a}{2} = 0.$$

Problem 39. Proof 1. Let $O = 0$:

$$(I - G)(B - C) = \left(\frac{aA + bB + cC}{a+b+c} - \frac{A+B+C}{3}\right)(B - C).$$

The numerator is

$$(a - 2b + c)B^2 + (-a - b + 2c)C^2 + (-2a + b + c)AB$$
$$+ (3b - 3c)BC + (2a - b - c)CA.$$

Using $B^2 = C^2 = R^2$, $2AB = 2R^2 - c^2$, $2BC = 2R^2 - a^2$, and $2CA = 2R^2 - b^2$, we get

$$\frac{1}{2}(c-b)(3a - b - c)(a + b + c) = 0.$$

So, $IU \perp BC$ if and only if $b = c$ or $b + c = 3a$.

Proof 2. Let $C = 0$:

$$(I - G)(B - C) = \left(\frac{aA + bB + cC}{a+b+c} - \frac{A+B+C}{3}\right)(B - C)$$

$$= \frac{(2a - b - c)AB + (-a + 2b - c)B^2}{3(a+b+c)}$$

$$= \frac{(2a - b - c)\frac{a^2+b^2-c^2}{2} + (-a + 2b - c)a^2}{3(a+b+c)}$$

$$= \frac{1}{6}(3a - b - c)(b - c).$$

So, $IU \perp BC$ if and only if $b = c$ or $b + c = 3a$.

Note. Proof 2 is simpler because the conclusion of the problem did not involve the circumcentre O. It was guessed that the condition involving O was added to exclude the case of an equilateral triangle ABC where $I - G$ would be zero. In specific calculations, $\left(\frac{aA+bB+cC}{a+b+c} - \frac{A+B+C}{3} \right)(B-C)$ has more occurrences of B and C than A. So, assuming $C = 0$ or $B = 0$ simplifies the calculations.

Problem 40. Let $D = I + k(A - B) = \frac{aA+bB+cC}{a+b+c} + k(A - B)$, where D lies on AC. Solving $\frac{b}{a+b+c} - k = 0$, we find $D = \frac{aA+bA+cC}{a+b+c}$.
Let $A = 0$, then

$$\left(0 - \frac{aA + bA + cC}{a+b+c} \right) \left(C - \frac{aA + bB + cC}{a+b+c} \right)$$

$$= \frac{(a+b+c)(a+b)OC - (a+b+c)bOB - (a+b)cC^2 + bcBC}{(a+b+c)^2}$$

$$= \frac{(a+b+c)(a+b)\frac{b^2}{2} - (a+b+c)b\frac{c^2}{2} - (a+b)cb^2 + bc\frac{b^2+c^2-a^2}{2}}{(a+b+c)^2}$$

$$= \frac{b(a+b)(b-c)}{2(a+b+c)}.$$

Hence, $AB = AC \iff OD \perp CI$.

Problem 41. $2|CE| = |CE| + |CG| = a+b+c$, $|AE| = |CE| - |CA| = \frac{a+b+c}{2} - b = \frac{a-b+c}{2}$,

$$E = \frac{\left(\frac{a-b+c}{2} + b \right) A - \frac{a-b+c}{2} C}{b} = \frac{(a+b+c)A - (a-b+c)C}{2b}.$$

Similarly,

$$G = \frac{(a+b+c)B - (-a+b+c)C}{2a}, \quad H = \frac{(a+b+c)B - (a+b-c)A}{2c},$$

$$I = \frac{(a+b+c)C - (a-b+c)A}{2b}, \quad F = \frac{(a+b+c)C - (-a+b+c)B}{2a},$$

$$D = \frac{(a+b+c)A - (a+b-c)B}{2c}, \quad X = \frac{E+H}{2}, \quad Y = \frac{D+I}{2}, \quad Z = \frac{G+F}{2}.$$

Computing, we get $\frac{A+Z}{2} = \frac{X+Y}{2}$.

Problem 42. Let $AB = 1$, $AD = t$, $D = (1-t)A + tB$, $E = (1-t)B + tC$,

$$F = \frac{(1-t)^2 A + t(1-t)B + t^2 C}{(1-t)^2 + t(1-t) + t^2}, \quad H = sF + (1-s)B,$$

$$G = 2H - E = \frac{2s(t-1)^2 A + (1 - 2s + 4st - 4st^2 + t^3)B - t(1 - t - 2st + t^2)C}{1 - t + t^2}.$$

Because G lies on CD, $\frac{2s(t-1)^2}{1-2s+4st-4st^2+t^3} = \frac{1-t}{t}$, solving for s yields $s = \frac{1+t}{2}$.

Let $A = 0$,

$$G^2 = \left(\frac{t(t-1)(1+t)B - t(2t-1)C}{1 - t + t^2} \right)^2$$

$$= t^2 \frac{(t-1)^2(1+t)^2 B^2 - (t-1)(1+t)(2t-1)BC + (2t-1)^2 C^2}{(1-t+t^2)^2}$$

$$= t^2 \frac{(t-1)^2(1+t)^2 - (t-1)(1+t)(2t-1) + (2t-1)^2}{(1-t+t^2)^2} = t^2.$$

Problem 43. Let $A = 0$, $I = \frac{aA+bB+cC}{a+b+c}$, $D = \frac{aA+cC}{a+c}$, $E = \frac{a\frac{bB+cC}{b+c}+cC}{a+c}$.

Let $F = tE + (1-t)I$, considering F is on AB, we get

$$t\left(1 - \frac{ab}{(a+c)(b+c)}\right) + \frac{(1-t)c}{a+b+c} = 0,$$

$$t = -\frac{(a+c)(b+c)}{a^2 + ab + b^2 + ac + bc},$$

$$F = \frac{b(b+c)B + a(a+b+c)A}{a^2 + ab + b^2 + ac + bc},$$

$$(F - D)I = \left(\frac{b(b+c)B + a(a+b+c)A}{a^2 + ab + b^2 + ac + bc} - \frac{aA + cC}{a+c} \right) \frac{aA + bB + cC}{a+b+c}$$

$$= \frac{b^2(a+c)(b+c)B^2 - bc(a^2 + b^2 - c^2)BC - c^2(a^2 + ab + b^2 + ac + bc)C^2}{(a+c)(a+b+c)(a^2 + ab + b^2 + ac + bc)}$$

$$= \frac{bc(a+b+c)(a^2 + b^2 - c^2)}{2(a+c)(a^2 + ab + b^2 + ac + bc)}.$$

So, $DF \perp AI \iff CA \perp CB$.

Note. After obtaining the coordinates of point F, it is observed that $DF \perp AI$ if and only if $|AD| = |AF|$. Hence,

$$|AD| - |AF| = \frac{bc}{a+c} - \frac{bc(b+c)}{a^2 + ab + b^2 + ac + bc}$$

$$= \frac{bc(a^2 + b^2 - c^2)}{(a+c)(a^2 + ab + b^2 + ac + bc)}.$$

Therefore, $DF \perp AI$ if and only if $CA \perp CB$.

Problem 44. Let $P = \frac{xA+yB+zC}{x+y+z}$ with x, y, z as positive numbers. Let $A_1 = \frac{yB+zC}{y+z}$, $B_1 = \frac{xA+zC}{x+z}$, and $C_1 = \frac{xA+yB}{x+y}$. If P is the centroid of $\triangle ABC$, then $x = y = z$, and hence $P = \frac{A_1+B_1+C_1}{3}$, implying that P is the centroid of $\triangle A_1 B_1 C_1$.

Conversely, if P is the centroid of $\triangle A_1 B_1 C_1$, then

$A_1 + B_1 + C_1 - 3P$

$= [x(y+z)(y^2 - x^2 + z^2 - yz)A + y(x+z)(x^2 - y^2 + z^2 - xz)B$

$+ (x+y)z(x^2 + y^2 - z^2 - xy)C]$

$(x+y)(x+z)(y+z)(x+y+z) = 0,$

which leads to the system of equations

$$\begin{cases} y^2 - x^2 + z^2 - yz = 0, \\ x^2 - y^2 + z^2 - xz = 0, \\ x^2 + y^2 - z^2 - xy = 0. \end{cases}$$

Summing these equations yields $x^2 + y^2 + z^2 - xy - yz - zx = 0$ or, equivalently, $(x-y)^2 + (y-z)^2 + (z-x)^2 = 0$. This implies $x = y = z$; therefore, $P = \frac{A+B+C}{3}$, proving that P is the centroid of $\triangle ABC$.

Problem 45. Proof 1. Let $\triangle ABC$ has circumcentre O at the origin, $K = -A$, and $I = \frac{aA+bB+cC}{a+b+c}$. $P = B+(A-B)\frac{\frac{a+b+c}{2}}{c}$ and $Q = C+(A-C)\frac{\frac{a+b+c}{2}}{b}$. Then,

$(K-I)(P-Q)$

$= \Big[-(b-c)(a+b+c)(2a+b+c)A^2 + b^2(a+b-c)B^2 + (-a+b-c)c^2C^2$

$+ 2abAB(a+b) + 2b^2cBC(1-bc) + 2acAC(-a-c) \Big] 2bc(a+b+c)$

$$= \Big[-(b-c)(a+b+c)(2a+b+c)R^2 + b^2(a+b-c)R^2 + (-a+b-c)c^2R^2$$
$$+ (2R^2 - c^2)ab(a+b) + (2R^2 - a^2)b^2c(1-bc) + (2R^2 - b^2)ac(-a-c)\Big]$$
$$\times 2bc(a+b+c)$$
$$= 0.$$

Proof 2. Let $A = 0$, $P = B + (A-B)\frac{a+b+c}{2c}$, $Q = C + (A-C)\frac{a+b+c}{2b}$, and $I = \frac{aA+bB+cC}{a+b+c}$. Then,

$$(K-I)(P-Q) = \Big[-(b-c)(a+b+c)(2a+b+c)A^2 + b^2(a+b-c)B^2$$
$$+ (-a+b-c)c^2C^2 + 2abAB(a+b) + 2b^2cBC(1-bc)$$
$$+ 2acAC(-a-c)\Big]2bc(a+b+c)$$
$$= \Big[-(b-c)(a+b+c)(2a+b+c)R^2 + b^2(a+b-c)R^2$$
$$+ (-a+b-c)c^2R^2 + (2R^2 - c^2)ab(a+b)$$
$$+ (2R^2 - a^2)b^2c(1-bc) + (2R^2 - b^2)ac(-a-c)\Big]$$
$$\times 2bc(a+b+c)$$
$$= 0,$$

$$(P-Q)(K-I)$$
$$= \left(\frac{(-a-b+c)B}{2c} - \frac{(-a+b-c)C}{2b}\right)\left(K - \frac{bB+cC}{a+b+c}\right)$$
$$= \frac{(b(-a-b+c)B - c(-a+b-c)C)((a+b+c)K - bB - cC)}{2bc(a+b+c)}.$$

The numerator part is
$$b(a+b+c)(-a-b+c)BK - c(a+b+c)(-a+b-c)CK$$
$$- b^2(-a-b+c)B^2 + bc(-a+b-c)BC$$
$$- bc(-a-b+c)BC + c^2(-a+b-c)C^2$$
$$= b(a+b+c)(-a-b+c)c^2 - c(a+b+c)(-a-b-c)b^2$$
$$- b^2(-a-b+c)c^2 + bc(-a+b-c)\frac{b^2+c^2-a^2}{2}$$
$$- bc(-a-b+c)\frac{b^2+c^2-a^2}{2} + c^2(-a+b-c)b^2$$
$$= 0.$$

Note. The choice of origin might affect the complexity of the solution. Proof 1 assumes the circumcentre as the origin, leading to the use of the circumradius R, which makes the solution slightly more complex. Proof 2 takes A as the origin and cleverly transforms BK into B^2 through projection, simplifying the computation.

Problem 46.

1. Let BM and CA intersect at point K. Due to the properties of the medians of an isosceles triangle, CM is the median of isosceles $\triangle BCK$. Since $\frac{\overrightarrow{CK}}{a} = \frac{\overrightarrow{CA}}{b}$, we have $K = \frac{a}{b}(A - C) + C$,

$$M = \frac{B+K}{2} = \frac{B + \frac{a}{b}(A-C) + C}{2} = \frac{aA + bB + (b-a)C}{2b},$$

$$D = \frac{\frac{a-b+c}{2}A + \frac{-a+b+c}{2}B}{c},$$

$$E = \frac{\frac{a+b-c}{2}A + \frac{-a+b+c}{2}C}{b}.$$

Since $\frac{-a+b+c}{c}M + \left(1 - \frac{-a+b+c}{c}\right)E = D$, points M, D, and E are collinear. By the same reasoning, point N also lies on line DE.

2. Let $A = 0$. Then, $M = \frac{aA + bB + (b-a)C}{2b}$, and due to symmetry,

$$N = \frac{aA + cC + (c-a)B}{2c},$$

$$(M-N)^2 = \frac{a^2(bB - cC)^2}{4b^2c^2} = \frac{a^2(b^2B^2 + c^2C^2 - 2bcBC)}{4b^2c^2}$$

$$= \frac{a^2(b^2c^2 + c^2b^2 - bc(b^2 + c^2 - a^2))}{4b^2c^2}$$

$$= \frac{a^2(a+b-c)(a-b+c)}{4bc},$$

$$\frac{a^2(a+b-c)(a-b+c)}{bc} - a^2 = \frac{a^2(a^2 - b^2 + bc - c^2)}{bc}.$$

So, $2MN = BC \iff \angle A = 60°$.

Note. The challenge in this problem is finding how to represent point M. In theory, M can be represented by three points: B, I, and C, and point I can be represented by three points: A, B, and C. However, without a careful technique, blindly calculating the foot of the perpendiculars will increase the workload.

A.3 Exercise 3.2

Problem 1. $2\left(\frac{P+A}{2} - \frac{C+A+C-B}{2}\right)(B-P) + [(C-P)^2 - (B-C)^2] = 0$.

Problem 2. $\left[(2A - B - \frac{A+B+C}{3})^2 - (A-C)^2\right] + \frac{32}{9}\left(A - \frac{B+C}{2}\right)B - \frac{A+C}{2} = 0$.

Problem 3. Let $B = 0$, then $\left(\frac{A+C}{2} - \frac{A}{3}\right)\left(\frac{A+C}{2} - \frac{2A}{3}\right) + \frac{1}{36}(A^2 - 9C^2) = 0$.

Problem 4. Let $A = 0$, then $-8BC + \left[9\left(\frac{2B+C}{3}\right)^2 - (2B-C)^2\right] = 0$.

Problem 5. Let $A = 0$, then $\left[\left(\frac{2B+C}{3}\right)^2 - \frac{4}{9}\left(B - \frac{C}{2}\right)^2\right] - \frac{8}{9}BC = 0$.

Problem 6. $4\left(\frac{A+D}{2} - \frac{B+C}{2}\right)^2 - (A-D+B-C)^2 + 4(A-C)(B-D) = 0$.

Note. When A and D coincide at one point, then M and O also coincide with A and D, and the above propositions become: the median of the hypotenuse of a right triangle is equal to half of the hypotenuse.

Problem 7. Let $D = 0$, then by $E - A = \frac{D-C}{2}$, we get $E = A - \frac{C}{2}$, and $\left(A - \frac{C}{2} - \frac{B}{2}\right)(B - C) + \frac{1}{2}((A-B)^2 - (A-C)^2) = 0$.

Problem 8. $4\left(\frac{A+D}{2} - \frac{B+C}{2}\right)\left(\frac{B+D}{2} - \frac{A+C}{2}\right) + (A-B)^2 - (C-D)^2 = 0$.

A.4 Exercise 3.3

Problem 1. $(A-B)^2 - (B-C)(A-D) + (A-C)(B-D) - (B-C)(B-A) - (B-A)(D-A) = 0$.

Problem 2. Let $A = 0$, then $[(B-C)^2 - 2B(B-D)] + (B^2 - C^2) - 2B(D-C) = 0$.

Problem 3. Let $A = 0$, then $(D-E)(B-C) - (EC+BD) + (CD+BE) = 0$.

Problem 4. Let $E = 0$, then $A(P-Q) - B(C-Q) + C(B-P) + Q(A-B) - P(A-C) = 0$.

Problem 5. Let $O = 0$, then $\left[\frac{2(A+B+C)}{3} - A\right]^2 + \left[\frac{2(A+B+C)}{3} - B\right]^2 + \left[\frac{2(A+B+C)}{3} - C\right]^2 - A^2 - B^2 - C^2 = 0$.

Problem 6. $(E-F)(A-B) - (D-A)(B-E) + [(B-A)(B-F) - (B-D)(B-E)] = 0$.

Note. Surprisingly, point C does not appear in the identity. However, upon careful consideration, it is indeed superfluous.

Problem 7. Let $C = 0$, then
$$6\left(2A + \frac{B}{2}\right)\frac{A-B}{3} + 3AB - (4A^2 - B^2) = 0.$$

Problem 8. Let $C = 0$, then $D = A - B$,
$$(A - B - B)\left[A - B - \left(2A - \frac{B}{2}\right)\right] + (A^2 - B^2) - \frac{3}{2}AB = 0.$$

Problem 9. Let $O = 0$, then $\left[4\left(\frac{A}{2} - \frac{C+B}{2}\right)^2 - (B^2 + C^2 - A^2)\right] - 2(A - B)(A - C) = 0.$

Problem 10. Let $H = 0$, then $(BE - \frac{A}{2}D) - B(E - A) + A(\frac{D}{2} - B) = 0.$

Problem 11. Let $F = 0$, then $[(A - B)^2 - A(A - E)] + 2B(A - \frac{B+C}{2}) + (BC - AE) = 0.$

Problem 12. Let $D = 0$, then $[(B - E)^2 - (F - C)^2] - [E^2 - (A - C)^2] + [F^2 - (A - B)^2] - 2AB + 2AC + 2BE - 2FC = 0.$

Problem 13. $[(B - D)^2 - (A - D)(B - C)] + (A - B)(B - C) - (D - B)(D - C) = 0.$

Problem 14. $4\left(\frac{A+D}{2} - C\right)\left(\frac{A+D}{2} - B\right) - [(A - B + D - C)^2 - (B - C)^2] = 0.$

Problem 15. $(A - C)^2 + (B - D)^2 - (A - B + D - C)^2 - 2(A - C)(B - D) = 0.$

Problem 16. The key to this problem lies in establishing a rectangle based on the perpendicular relationships. One observes that
$$E = \frac{A + D}{2} + C - \frac{A + B}{2}.$$
The expression
$$[(C - D)^2 - (C - B)^2] - 2(D - A)(D - B) + 2(E - A)(D - B) = 0.$$

Problem 17. The equation
$$(A - C)^2 + (B - D)^2 - 4\left(\frac{A+B}{2} - \frac{C+D}{2}\right)^2 + 2(A - C)(B - D) = 0.$$

Note. From the proof, it is evident that the proposition holds not necessarily for convex quadrilaterals but even for spatial quadrilaterals. The intersection point O is purely redundant.

Problem 18. $(A - D)^2 - (A - C)^2 - (B - D)^2 - 2\left(\frac{B+C}{2} - D\right)(A - B) + (A - C)(B - C) = 0.$

Problem 19. The expression
$$\left(F - \frac{A+C-B+C}{2}\right)\left(E - \frac{A+C-B+A}{2}\right)$$
$$-\left(\frac{A+B}{2} - E\right)\cdot\left(\frac{B+C}{2} - F\right)$$
$$+ 2\left(B - \frac{A+C}{2}\right)\left(\frac{A+C}{2} - \frac{E+F}{2}\right) = 0.$$

Problem 20. Let $H = 0$ and $K = \frac{A}{2} + N - A$. We have
$$\left(\frac{A}{2} - C\right)\left(\frac{A}{2} - N\right) + (A - N)C - \frac{1}{2}(C - K)A = 0.$$

Problem 21. $2\left(A - \frac{B+D}{2}\right)\left(C - \frac{A+D}{2}\right) + (A - B)(A - C) + (A - D)\left(\frac{B+D}{2} - C\right) = 0.$

Problem 22. Let $C = 0$. The equation becomes
$$\left(D - \frac{B+E}{2}\right)B - \left(A - \frac{B}{2}\right)(B+E) + [AE - B(D-A)] = 0.$$

Problem 23. The expression is
$$\left(\frac{A+K}{2} - B\right)\left(\frac{A+K}{2} - \frac{A+C-B+C}{2}\right) + \frac{1}{2}(B-A)(B-C)$$
$$- \frac{1}{2}(K-B)(K-C) - \frac{1}{4}(K-B)(A-K) = 0.$$

Problem 24. Expressing points B, C, H, and P as concyclic is the key to this problem. Let O be the circumcentre of $\triangle ABC$, and $H = A + B + C$. Based on $A^2 = B^2 = C^2$, we have $(B+C-(A+B+C))^2 = (B+C-C)^2 = (B+C-B)^2$. So, B, C, and H all lie on a circle with centre $B + C$ and radius $|OA|$. Additionally, P lies on this circle, so $(B+C-P)^2 - A^2 = 0$:
$$2\left(A - \frac{B+C}{2}\right)(A+B+C-P) - (A-P)(A+B+C-P)$$
$$+ [(B+C-P)^2 - A^2] = 0.$$

Problem 25. The equation is
$$\left(\frac{C+E}{2} - B\right)\left(\frac{C+E}{2} - A\right) + \frac{1}{2}(B-E)\left(C - \frac{2A+B}{3}\right) + \frac{1}{4}(B-E)$$
$$\left(E - \frac{2A+B}{3}\right) + \frac{1}{4}((A-B)^2 - (A-C)^2) = 0.$$

Note. Take the midpoint G of EB, and construct the parallelogram $GFAH$. Then H lies on the line CD, and G is the orthocentre of $\triangle HBF$.

Problem 26. Let $P = 0$, the equation becomes
$$B(A - C) - D(B - C) - G(A - B)$$
$$+ [(B - A)(B - G) - (B - D)(B - C)] = 0.$$

Problem 27. Let $O = 0$, then
$$-D(A - C) - B(B - A) + C(C - D) + (B^2 - C^2) + A(D - B) = 0.$$

Problem 28. The equation is $\left(\frac{A+B}{2} - \frac{2A+(A+C-B)}{3}\right)\left(\frac{11C-5B}{6} - \frac{2A+(A+C-B)}{3}\right) + \frac{5}{12}(B - A)(B - C) - \frac{1}{2}((B - A)^2 - (B - C)^2) = 0.$

Problem 29. Let $A = 0$,
$$\left(\frac{B+C}{2} - E\right)\left(\frac{B+C}{2} - F\right) - \frac{B+C}{2}\left(\frac{B+C}{2} - (E+F)\right) - EF = 0.$$

Problem 30. Let $A = 0$. The equation is
$$(E - G)(F - G) - EF + G(E + F - G) = 0.$$

Problem 31. Let $B = 0$, then
$$8\left(\frac{A}{2} - (A + C)\right)\left(\frac{A}{2} - \frac{C}{4}\right) + 3AC + 2(A^2 - C^2) = 0.$$

Problem 32.
$$4\left(\frac{A+B+C+D}{4} - A\right)\left(\frac{A+B+C+D}{4} - D\right) + (A - D)^2$$
$$- \left(\frac{A+D}{2} - \frac{B+C}{2}\right)^2 = 0.$$

Problem 33. Let $P = 0$:
$$2O(C - D) - 2C\left(O - \frac{A}{2}\right) + 2D\left(O - \frac{B}{2}\right) + (BD - AC) = 0.$$

Problem 34. Let $P = 0$, then
$$-2A\left(Q - \frac{C}{2}\right) + 2B\left(Q - \frac{D}{2}\right) + 2Q(A - B) + (BD - AC) = 0.$$

So, Q lies on the perpendicular bisector of CP and is the circumcentre of $\triangle CDP$.

Problem 35.

$$\left(D - \frac{B+C}{2}\right)\left(D - \frac{B+A}{2}\right)$$

$$- \frac{1}{4}(B-C)(B-A) - (D-B)\left(D - \frac{A+C}{2}\right) = 0.$$

Problem 36.

$$(C-G)(B-D) - \frac{1}{2}(C-A)(B-D) + \left(G - \frac{A+D}{2}\right)(B-C)$$

$$- \left(G - \frac{A+B}{2}\right)(D-C) = 0.$$

Problem 37. $(A-B)^2 + (C-D)^2 + (E-F)^2 - (B-C)^2 - (D-E)^2 - (F-A)^2 + 2[(A-G)(B-F) + (C-G)(D-B) + (E-G)(F-D)] = 0.$

An alternative proof. Repeatedly use the Pythagorean theorem:

$$(AB^2 - FA^2) + (CD^2 - BC^2) + (EF^2 - DE^2)$$
$$= (GB^2 - FG^2) + (GD^2 - BG^2) + (GF^2 - DG^2) = 0.$$

Problem 38. Let the circumcentre of $\triangle BDP$ be $O = 0$, then

$$\left[(B-A)(B-P) - 2\left(B - \frac{A+C}{2}\right)^2\right] + 2B(B-C) + 2\frac{B+P}{2}(B-A)$$

$$+ 2\left[\left(\frac{A+C}{2}\right)^2 - B^2\right] = 0.$$

Problem 39. Let $\frac{B+C}{2} = 0$, then $C = -B$, and $K = A - D$,

$$2A\left(A - D - \frac{A+N}{2}\right) + [(D-B)(D-(-B)) - (D-A)^2]$$

$$+ [AN - B(-B)] = 0.$$

Problem 40. Let $O = 0$. The equation is

$$(A-B)^2 + (A-C)^2 - 2(C-B)^2 + 12\left(\frac{A+B+C}{3}\right)\left(A - \frac{B+C}{2}\right)$$

$$+ 3(B^2 + C^2 - 2A^2) = 0.$$

Problem 41.

$$\frac{3}{2}\left(A - \frac{C+2K}{3}\right)(A-B+C-K)$$

$$+ \left[(A-B)^2 - \left(C - \frac{A+C-B+K}{2}\right)^2\right]$$

$$+ \frac{3}{4}[(A-C)^2 - (2A-B-K)^2] = 0.$$

A.5   Exercise 4.1
Problem 1.

$$\left[\left(W - \frac{2W+C}{3}\right)^2 - (W-A)^2\right]$$

$$+ 2\left(\frac{2W+C}{3} - A\right)\left(\frac{2W+C}{3} - \frac{A+C}{2}\right) = 0.$$

Problem 2.

$$\left(\frac{A+C}{2} - \frac{A+B}{2}\right)\left(\frac{A+C}{2} - \frac{2C-A+B}{2}\right)$$

$$+ \frac{1}{2}\left(A - \frac{B+C}{2}\right)(B-C) = 0.$$

Problem 3. Let $P = 0$, then

$$Q(A-B) + \frac{1}{2}(BD - AC) + \left(Q - \frac{D}{2}\right)B - \left(Q - \frac{C}{2}\right)A = 0.$$

Problem 4. Let $D = 0$, then

$$2AB + 2\frac{B+C}{2}\left[C - \left(\frac{B+C}{2} + A\right)\right] + \left(\frac{B+C}{2} - A\right)(B-C) = 0.$$

Problem 5.

$$[(E-G)^2 - (E-D)(E-F)] + (F-A)(D-B) - (G-A)(G-B)$$
$$- (E-A)(D-G) + (E-B)(G-F) = 0.$$

Problem 6. Let $H = 0$, then

$$[(A-P)^2 - (B-Q)^2] - (P^2 - Q^2) - 2A\left(\frac{B+A}{2} - P\right)$$

$$+ 2B\left(\frac{B+A}{2} - Q\right) = 0.$$

Similarly to prove $AP = CR$.

Problem 7. Let the circumcentre of $\triangle ABC$ be the origin, and $H = A + B + C$:

$$(A+B+C-F)(A-F) - 2\left(\frac{B+C}{2} - \frac{E+F}{2}\right)(A-F)$$

$$- 2\frac{A+E}{2}(A-F) = 0.$$

Problem 8. Let the circumcentre O of $\triangle ABC$ be the origin, and $H = A + B + C$. It is easy to verify that the circumcentre of $\triangle BHC$ is $B + C$, and the square of the radius is A^2. The original proposition is equivalent to proving $XM \perp XH$:

$$2\left(X - \frac{C+B}{2}\right)(X - (A+B+C)) - (X-A)(X-(A+B+C))$$

$$+ [A^2 - (B+C-X)^2] = 0.$$

Problem 9.

$$(A-F)(A-G) - (A-B)(C-G) - (A-C)(B-F)$$

$$- [(A-B)(A-C) + (B-F)(C-G)] = 0.$$

Another Proof. $\overrightarrow{AF} \cdot \overrightarrow{AG} = (\overrightarrow{AB} + \overrightarrow{BF}) \cdot (\overrightarrow{AC} + \overrightarrow{CG}) = \overrightarrow{AB} \cdot \overrightarrow{AC} + \overrightarrow{BF} \cdot \overrightarrow{CG} = 0$, so $AF \perp AG$.

Problem 10. Analysis: Let $\triangle ABC$'s circumcentre be $O = 0$. The equation $\sin^2 A + \sin^2 B + \sin^2 C = 1$ is equivalent to $a^2 + b^2 + c^2 = 4R^2$, i.e., $(A-B)^2 + (B-C)^2 + (C-A)^2 = 4A^2$. The nine-point circle centre $O_9 = \frac{A+B+C}{2}$ with radius squared $R_9^2 = \left(\frac{A}{2}\right)^2$. To prove that the circles are orthogonal, it suffices to show $R_9^2 + R^2 = OO_9^2$, i.e., $\left(\frac{A}{2}\right)^2 + A^2 = \left(\frac{A+B+C}{2}\right)^2$.

Identity: Let $O = 0$:

$$4\left[\left(\frac{A}{2}\right)^2 + A^2 - \left(\frac{A+B+C}{2}\right)^2\right] - 3(2A^2 - B^2 - C^2)$$
$$- [(A-B)^2 + (B-C)^2 + (C-A)^2 - 4A^2] = 0.$$

Problem 11. Let $H = 0$. The equation is

$$4\frac{E+D}{2}\left[\frac{E+D}{2} - \left(B + D - \frac{B+C}{2}\right)\right]$$
$$+ (B+D)(D-C) - (C+E)(E-B) = 0.$$

Problem 12. Let $O = 0$:

$$\left(A + \frac{B+C}{2}\right)\left(A - \frac{E+C}{2}\right)$$
$$- \frac{1}{4}(C-B)(C-E) - \frac{A+B}{2}(A-E) - \frac{1}{2}(A^2 - C^2) = 0.$$

Problem 13. Let $O = 0$:

$$(P-C)(A_1 - B_1) + (P-A)(B_1 - C_1) + (P-B)(C_1 - A_1)$$
$$- A_1(B-C) - B_1(C-A) - C_1(A-B) = 0.$$

Problem 14. Let $A = 0$:

$$12\left(\frac{B+C}{2} - \frac{C+K}{3}\right)\left(\frac{B+C}{2} - P\right)$$
$$- 2(PC - BK) + 3[P^2 - (B-P)^2] + [K^2 - (C-K)^2]$$
$$+ 4[(B-P)(K-C) - BK] = 0.$$

Problem 15. The equation is

$$\left(\frac{A+B}{2} - M\right)\left(\frac{A+B}{2} - D\right) - (B-M)(A-D)$$
$$- \frac{1}{2}\left(M - \frac{B+D-A+D}{2}\right)(A-B) = 0.$$

Note.

$$\left(M - \frac{C+D}{2}\right)(C-D) = \left(M - \frac{B+D-A+D}{2}\right)(A-B).$$

Problem 16. $(A-G)(C-H) - (C-G)(A-B) - (E-A)(H-B) - (A-C)(E-B) + 2(E-G)\left(\frac{A+H}{2} - \frac{B+C}{2}\right) = 0.$

Problem 17.

$$4\left(C - \frac{B+A}{2}\right)\left(D - \frac{B+C}{2}\right) - 2[(D-A)^2 - (D-C)^2]$$
$$+ 2(A-D)(A-B) + (B-C)(A-B) = 0.$$

Problem 18.

$$3(A-B)(B-C) + 4\left(A - \frac{A+C-B+C}{2}\right)\left(B - \frac{A+C-B+A}{2}\right)$$
$$+ 2[(A-B)^2 - (B-C)^2] = 0.$$

Problem 19. $B_1 = A_1 + B - A, C_1 = A_1 + C - A$, $[(A-B)^2 - (A-C)^2] - (A-A_1)(A-B) - (A-A_1)(A-C) + (A-B_1)(B-C_1) - (B-C_1)(C-A_1) = 0$. Hence, $(A-B)^2 = (A-C)^2$, and similarly, $(A-B)^2 = (B-C)^2$. Therefore, the prism is a regular triangular prism.

Problem 20. Pass through the midpoint O of AB to make a perpendicular segment OM to CD. It is easy to see that $\triangle MOC \sim \triangle ECD$, and hence $\frac{OC}{CM} = \frac{CD}{DE}$, implying $\frac{AB}{CD} = \frac{CD}{DE}$. Thus, $AB^2 = \frac{AB}{DE} \cdot CD^2$.

$$[(A-E)^2 + (B-E)^2 - (A-B)^2 - 2(E-C)^2] - 2(C-A)(C-B)$$
$$+ 4\left(C - \frac{A+B}{2}\right)(C-E) = 0.$$

Problem 21. $(A-B)^2 + (C-D)^2 + (E-F)^2 - (B-C)^2 - (D-E)^2 - (F-A)^2 + 2[(A-C)(B-G) + (C-E)(D-G) + (E-A)(F-G)] = 0.$

Note. From the identity, it can be observed that the conditions "$AB = AF$, $BC = CD$, $DE = EF$" can be weakened to $AB^2 + CD^2 + EF^2 = BC^2 + DE^2 + FA^2$.

Problem 22.

$$2\left(\frac{X+Y}{2} - M\right)(A-B) + 2\left(\frac{Y+Z}{2} - M\right)(B-C)$$
$$+ 2\left(\frac{Z+X}{2} - M\right)(C-A)$$
$$- (P-X)(B-C) - (P-Y)(C-A) - (P-Z)(A-B) = 0.$$

Note 1. This problem appears complex at first, but it possesses strong symmetry (note that the expressions such as $A - B$, $B - C$, and $C - A$ exhibit symmetry). In fact, once the six polynomials are written down, the identity can be directly obtained through observation, avoiding the need to solve the equation system step by step.

Note 2. The identity indicates that point X does not necessarily need to lie on BC; other points can be further expanded as well. Similar issues exist in the other problems, and they are not explicitly mentioned one by one.

Problem 23.

Given. $O = 0, H = A + B + C, X = B + C, Y = A + C, Z = B + A$,
$$P = \frac{A+B+C}{2} = \frac{0+(A+B+C)}{2} = \frac{A+(B+C)}{2}$$
$$= \frac{B+(A+C)}{2} = \frac{(A+B)+C}{2},$$
$$\left(\frac{A+B+C}{2} - \frac{B+C}{2}\right)^2 = \left(\frac{A+B+C}{2} - \frac{C+A}{2}\right)^2$$
$$= \left(\frac{A+B+C}{2} - \frac{A+B}{2}\right)^2 = \frac{R^2}{4},$$
where R is the circumradius of $\triangle ABC$.

Problem 24. $2\left(\frac{A}{2} - \frac{C+B}{2}\right)(H1 - H2) + A(B - C) + (C - H1)(A - B) - (B - H2)(A - C) - (B - H1)C + (C - H2)B = 0$.

Note. Observing that D, E, and F are redundant, the problem can be simplified.

Problem 25. Let $O = 0$, $2D(B-C) + [(B-D)(B-C) - (B-F)(B-A)] + 2\frac{A+E}{2}(C-A) - 2\frac{A+F}{2}(B-A) - [(C-D)(C-B) - (C-E)(C-A)] = 0$.

Problem 26.
$$(2C - A - E)(2C - A - A) - 4(C - A)(C - B)$$
$$- 2(A - E)(D - B) + 2(B - E)(A - B)$$
$$- 2(B - E)(C - D) - 4\left(\frac{A+B}{2} - \frac{C+D}{2}\right)(A - B) = 0.$$

Appendix: Exercise Answers

Problem 27. Let $D = 0$, then

$$FC - BE + A(B - C) - A(F - E)$$
$$+ [(A - B)(A - E) - (A - C)(A - F)] = 0.$$

Note. D is not necessarily outside $\triangle ABC$; it might not even lie on the plane of ABC.

Problem 28.

$$[(F - C)(C - D) - (A - B)(D - B)]$$
$$+ 2\left[\left(\frac{B+C}{2} - E\right)\left(\frac{B+C}{2} - A\right) - \left(\frac{B+C}{2} - B\right)\left(\frac{B+C}{2} - C\right)\right]$$
$$+ (A - B)(D - F) + (A - D)(C - F) + 2\left(\frac{B+C}{2} - A\right)(E - F) = 0.$$

Problem 29. Let $A = 0$, then

$$[(F - E)(G - H) - E(C - G)] - \left(F - \frac{B}{2}\right)G - \left(C - \frac{B}{2}\right)(C - E)$$
$$+ C(C - B) + \frac{1}{2}BE - \frac{1}{2}(G - C)B - EH + FH = 0.$$

Problem 30.

$$(D - B)^2 - (D - C)^2 - 3[(H - B)^2 - (H - C)^2]$$
$$- 6\left(H - \frac{A+B+C}{3}\right)(B - C) + 2(D - A)(B - C) = 0.$$

Problem 31.

$$4(P - Q)\left(\frac{A+C}{2} - \frac{B+D}{2}\right) + [(P - A)^2 - (P - B)^2]$$
$$+ [(P - C)^2 - (P - D)^2]$$
$$- [(Q - A)^2 - (Q - D)^2] + [(Q - B)^2 - (Q - C)^2] = 0.$$

Problem 32. Analysis. Using the identical method. assuming $M1 = 2N - C$, prove that $M1D \perp BE$.

Proof. Assume $A = 0, D = tB, E = tC$:

$$(2N - C - D)(B - E) - 2N(B - E) + t(B^2 - C^2) + (1 - t^2)BC = 0.$$

Problem 33.
$$O = 0, [(A-B)^2 - (C-D)^2] - (A+B+C+D)^2$$
$$+ [(A+B+C)^2 - D^2] + [(A+B+D)^2 - C^2]$$
$$- 2(A^2 + B^2 - C^2 - D^2) = 0.$$

Problem 34. Let the circumcentre of $\triangle ABC$ be O:
$$2\left(C - \frac{B+A+B+C}{2}\right)(A-K) - (C-K)(A-K)$$
$$+ [(A+B-K)^2 - (A+B-A)^2] = 0.$$

Note. The key observation is that the circumcentre of $\triangle ABH$ is $A + B$.

Problem 35. Let the circumcentre of $\triangle ABC$ be $O = 0$, and $H = A + B + C$. It is easy to see that the circumcentre of $\triangle ABH$ is $A+B$. To prove that C, K, and X are collinear is equivalent to proving that $KA \perp KX$:
$$2(K-A)\left(K - \frac{A+B+C+B}{2}\right) - (K-A)(K-C)$$
$$+ [B^2 - (K - (B+A))^2] = 0.$$

Problem 36. Let $A = 0$, then
$$[(C-K)(C-H) - (B-K)(B-L) - (C^2 - B^2)] - K(B-C)$$
$$+ K(L-H) + (F-L)(B-C) + [(C-B)F - C(L-H)] = 0.$$

Note. $(C - B)F = C(L - H)$ is used from the similarity of $\triangle AFE$ and $\triangle ACB$.

Problem 37.
$$4\left(P - \frac{A+C}{2}\right)(A-Q) + [(P-A)^2 - (P-B)^2]$$
$$- 2\left(P - \frac{B+C}{2}\right)(A-D) + [(Q-A)^2 - (Q-B)^2]$$
$$+ 2\left(Q - \frac{A+D}{2}\right)(B-C) - 4\left(Q - \frac{B+D}{2}\right)(B-P) = 0.$$

Problem 38. Let $H = 0, Q = B + C - P$:
$$2\left(\frac{B+C}{2} - A\right)(B+C-P) + B(A-C) + C(A-B)$$
$$+ 2\left[\left(\frac{B+C}{2} - B\right)\left(\frac{B+C}{2} - C\right) - \left(\frac{B+C}{2} - A\right)\left(\frac{B+C}{2} - P\right)\right] = 0.$$

Problem 39. Let the circumcentre of $\triangle ABC$ be at the origin:

$$[(P-E)^2 - (P-F)^2] - [(B-D)^2 - (B-F)^2]$$
$$+ [(C-D)^2 - (C-E)^2] - 2(D-A)(B-C) + 4\frac{P+A}{2}(E-F)$$
$$- 4\frac{A+B}{2}(A-F) + 4\frac{A+C}{2}(A-E) = 0.$$

Note. Using $\frac{P+A}{2}(E-F) = 0$ represents $AP \parallel EF$.

A.6 Exercise 4.2

Problem 1.

$$2\left(\frac{B+C}{2} - \frac{D+A}{2}\right)(O_1 - O_2)$$
$$+ \frac{1}{2}[(A-C)^2 - (B-D)^2] + \left(O_2 - \frac{C+D}{2}\right)(C-D)$$
$$+ \left(O_1 - \frac{A+P}{2}\right)(A-C) + \left(O_1 - \frac{B+P}{2}\right)(D-B)$$
$$+ \left(O_2 - \frac{C+P}{2}\right)(P-A) + \left(O_2 - \frac{D+P}{2}\right)(B-P) = 0.$$

Problem 2.

$$(B-C)(D-K) + (E-C)\left(K - \frac{A+E}{2}\right) + (F-B)\left(\frac{A+F}{2} - K\right)$$
$$+ (E-F)\left(K - \frac{E+F}{2}\right) - (D-F)(B-O)$$
$$+ \left(\frac{A+B}{2} - F\right)\left(\frac{A+B}{2} - O\right)$$
$$+ (D-E)(C-O) + (A-E)\left(O - \frac{A+C}{2}\right)$$
$$+ \left(C - \frac{B+C}{2}\right)\left(\frac{B+C}{2} - O\right)$$
$$+ \frac{1}{2}(A-C)\left(\frac{A+C}{2} - O\right) = 0.$$

Problem 3.

$$[(A - A_1)^2 + (B - B_1)^2 - (C - C_1)^2 - (D - D_1)^2]$$
$$- 2(A_1 - A)(A_1 - P) - 2(B_1 - B)(B_1 - P) + 2(C_1 - C)(C_1 - P)$$
$$+ 2(D_1 - D)(D_1 - P) + (C - A)(C - B) + (D - A)(D - B)$$
$$- 4\left(\frac{A+B}{2} - \frac{C+D}{2}\right)\left(\frac{A+B}{2} - P\right)$$
$$+ [(P - A_1)^2 + (P - B_1)^2 - (P - C_1)^2 - (P - D_1)^2] = 0.$$

Problem 4.

Because $\angle MEA = \angle EMO = \angle OPM$, so $EA \parallel PO$.
To prove that EF is the tangent to circle A, we just need to prove $OP \perp EF$:

$$2(O - P)(E - F) - [(P - E)(P - M) - (P - F)(P - N)]$$
$$+ 2\left(O - \frac{M+P}{2}\right)(P - E) - 2\left(O - \frac{N+P}{2}\right)(P - F) = 0.$$

Problem 5.

$$[(M - O)(A - B) - (M - O)(E - D)]$$
$$+ [(K - O)(E - F) - (K - O)(C - B)]$$
$$+ [(N - O)(C - D) - (N - O)(A - F)]$$
$$+ [(K - E)(K - B) - (K - F)(K - C)]$$
$$+ [(N - C)(N - F) - (N - D)(N - A)]$$
$$+ [(M - A)(M - D) - (M - B)(M - E)] = 0.$$

Problem 6.

$$(E - F)(G - H) + [(E - A + C - F)(B - A)$$
$$- (B - H + G - (A + C - B))(B - C)]$$
$$+ (C - F + E - B)(H - B) + (G - A)(B - E + F - C)$$
$$+ [(C - B)^2 - (A - B)^2] + (A - B)(C - B) = 0.$$

Note. First, consider how to transform the condition $BE + BH + DF + DG = 2$. It easily transforms into $EA + CF = BH + DG$. If it is further transformed into $(E - A + C - F)^2 = (B - H + G - (A + C - B))^2$, it will generate many square terms, making cancellation cumbersome. We transform it into

$$(E - A + C - F)(B - A) = (B - H + G - (A + C - B))(B - C).$$

Problem 7. Let $\triangle ABC$ have the orthocentre as the origin:

$$(I - D)(F - E) + 2\left(\frac{I}{2} - A\right)(B - C) + 2\left(E - \frac{B+D}{2}\right)\left(B - \frac{C+D}{2}\right)$$

$$- 2\left(F - \frac{C+D}{2}\right)\left(C - \frac{B+D}{2}\right) - C(A - E) + B(A - F)$$

$$- 2\left[(B - A)\left(B - \frac{B+D}{2}\right) - (B - E)\left(B - \frac{I}{2}\right)\right]$$

$$+ 2\left[(F - C)\left(A - \frac{I}{2}\right) - (A - C)\left(F - \frac{C+D}{2}\right)\right] = 0.$$

Note. According to the properties of the orthocentre, I is symmetric to the orthocentre about BC, and $\frac{I}{2}$ is the intersection point of AI and BC. Triangles $\triangle BE\frac{B+D}{2} \sim \triangle BA\frac{I}{2}$ and $\triangle CF\frac{C+D}{2} \sim \triangle CA\frac{I}{2}$ yield the last two relationships.

Problem 8. $4\left(\frac{A+B}{2} - \frac{C+D}{2}\right)(M - N) + (D - A)(D - B) - (C - A)(C - B) + 2(M - A)(A - B) + 2(N - B)(A - B) - 4\left(M - \frac{A+D}{2}\right)\left(A - \frac{C+D}{2}\right) + 4\left(N - \frac{B+C}{2}\right)\left(B - \frac{C+D}{2}\right) = 0.$

Note. It is inconvenient to express in parallel, so we prove it by showing orthogonality.

Problem 9. $(A - C)(P - C) - (A - P)(B - C) - (A - B)(A - H) + 2\left(\frac{A+B}{2} - C\right)(H - P) + (A - H)^2 - (C - H)^2 = 0.$

Note. The critical step in this problem involves expressing the point P on the line HI. Let K be the intersection point of lines AB and CF. The relations $(K - C)(K - F) = (K - A)^2 = (K - B)^2$ imply that K is the midpoint of AB. Since CF is perpendicular to HI, it follows that $\left(\frac{A+B}{2} - C\right)(H - P) = 0.$

Problem 10. $(E - F)(B - C) - [(E - A)(E - C) - (E - B)(E - D)] + (F - A)(E - C) - (F - D)(E - B) = 0.$

Problem 11. Let $O = 0$, then $2A(D - M) - [D^2 - (D - P)(D - A) - M^2] + (E - A)(P - E) - (A - P)(D - E) + [(A - E)^2 - (A - M)^2] = 0.$

Problem 12. $(A-E)(D-K)+(K-A)(K-E)+(A-D)(B-E)-(C-A)(C-B)-(D-C)(A-B)+\left[(A-C)^2-(A-K)^2\right]=0.$

Problem 13. Let $O=0$ and AC intersect BD at $K=E+F-P$. Then,

$$\begin{aligned}
&2P(E-F)+[(P-A)(P-B)-(P-C)(P-D)]\\
&\quad -2\left[(P-A)(P-F)-(P-E)(P-D)\right]\\
&\quad +[(D-P)(D-(E+F-P))-(D-F)(D-C)]\\
&\quad -[(A-P)(A-(E+F-P))-(A-E)(A-B)]\\
&\quad -2\frac{A+C}{2}(P-F)+2\frac{B+D}{2}(P-E)=0.
\end{aligned}$$

Note. The solution involves the concylic of points A, B, C, and D. Notably, $\triangle BAK \sim \triangle BPF$, $\triangle PEA \sim \triangle PFD$, $\triangle CKD \sim \triangle PFD$, $\triangle PEA \sim \triangle BKA$, $AC \parallel PF$, and $BD \parallel PE$.

Problem 14. Let $C=0$, then

$$\begin{aligned}
&\left[(A-P)^2+(B-Q)^2-P^2-Q^2\right]-2AB+2AQ+2PB\\
&\quad +2(P-D)(A-B)-2\left(Q-\left(D-\frac{A-B}{2}\right)\right)(A-B)=0.
\end{aligned}$$

Note. Point D is not necessarily on AB or even on the plane of $\triangle ABC$.

Problem 15. In Figure A.1,

$$\begin{aligned}
&[(E-B)^2-(C-D)^2]+[(A-C)^2-(B-C)^2]\\
&\quad +\left(\frac{A+B+C}{3}-(A+C-D)\right)^2\\
&\quad -\left(\frac{A+B+C}{3}-E\right)^2\\
&\quad +4\left(\frac{A+B+C}{3}-\frac{A+C-D+A}{2}\right)(E-C)\\
&\quad -4\left(\frac{A+B+C}{3}-\frac{B+E}{2}\right)(A-D)=0.
\end{aligned}$$

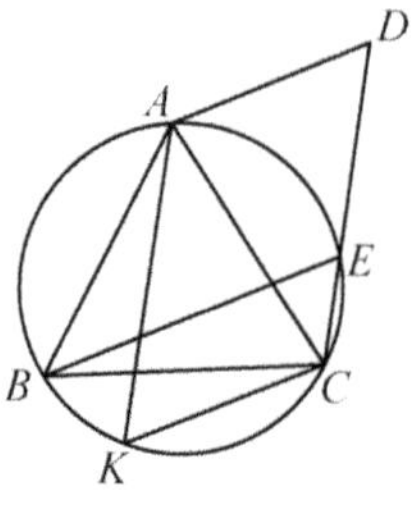

Figure A.1

Note. Generating this identity was not straightforward; refer to the diagram on the right. To some extent, auxiliary lines were added. Additionally, this problem utilizes area (exterior product) to construct the identity. Note that $\angle D = \angle BEC$, and $\angle DAC + \angle BCE = 180°$. Using the law of cosines, we have $\frac{|AD||CD|}{|BE||CE|} = \frac{S_{\triangle ADC}}{S_{\triangle CEB}} = \frac{|AD||AC|}{|CE||BC|}$, which simplifies to $\frac{|CD|}{|BE|} = 1$, or proves $\triangle ABE \cong \triangle ACD$.

Problem 16. Let $H = 0$ and $D = A - \frac{M-N}{2}$:

$$\frac{M+N}{2}\left(C - \left(A - \frac{M-N}{2}\right)\right) + B(A - C)$$

$$-\frac{1}{2}A(B - N) + \frac{1}{2}C(B - M)$$

$$+ \frac{1}{4}[(B - M)(B - (2A - M)) - (B - N)(B - (2C - N))] = 0.$$

Note. Construct the symmetric point A_1 of A about M and the symmetric point C_1 of C about N. Due to $\angle A = \angle C$, we have $\angle DMN = \angle A_1C_1B$, so points M, A_1, C_1, and N are concyclic.

Problem 17.

$$[(P - A)^2 - (P - Q)^2] + \left[2\left(\frac{A+P}{2} - C\right)(C - O_2)\right.$$

$$\left. - (P - A)(O_2 - B)\right] - (A - P)(A - B) + 2(P - Q)(O_2 - Q)$$

$$+ [(O_2 - C)^2 - (O_2 - Q)^2] + (C - A)(C - P) = 0.$$

Note. We take O_2 as the origin and ignore the points C and Q. Let R and r denote the radii of the large and small circles, respectively. The equation can be simplified to

$$((A - P)^2 - (P - Q)^2) + ((P - Q)^2 - P^2 + R^2)$$
$$+ (P^2 - (A - B)^2 - (2r - R)^2)$$
$$+ ((A - B)^2 - (R + r)^2 + (R - r)^2) = 0.$$

Problem 18. Let $O = 0$, then $(B^2 - C^2) + [(B - \frac{B+C}{2})(B - C) - (B - F)(B - A)] - [(C - \frac{B+C}{2})(C - B) - (C - E)(C - A)] - 2\frac{A+E}{2}(A - C) + 2\frac{A+F}{2}(A - B) = 0.$

Problem 19.

Proof 1. Let $H = 0$, then $2D\left(\frac{F+C}{2} - \frac{E+B}{2}\right) + (D - C)(D - F) - (D - B)(D - E) + C(B - F) - B(C - E) = 0.$

Proof 2. Let $D = 0$, then $2H\left(\frac{F+C}{2} - \frac{E+B}{2}\right) + (BE - CF) - (C - H)(B - F) + (B - H)(C - E) = 0.$

Problem 20. Suppose $H = 0$, then

$$2\left[\left(\frac{A+B}{2} - E\right)^2 - \frac{A+B}{2}\left(\frac{A+B}{2} - Q\right)\right]$$
$$+ A(B-C) + B(A-C) - 2(B-E)(A-E) + 2(C-Q)\frac{A+B}{2} = 0.$$

Problem 21. Let $O = 0$, then $3\left(E - \frac{B+2C}{3}\right)(H' - F) - (E - C)(A - F) - 2\left(E - \frac{B+C}{2}\right)(B - F) + (B - C)(B + C - E) + (E - C)(A + B + C - H') + 2H'\left(\frac{B+C}{2} - E\right) = 0.$

Problem 22. Let the circumcentre of $\triangle ABC$ be the origin, and $H = A + B + C$. It is evident that B, C, and H lie on the circle with centre $B + C$ and radius equal to the circumradius of $\triangle ABC$. The following identity shows that M also lies on this circle:

$$(B+C-M)^2 - (B+C-B)^2 + \frac{1}{2}\left[(B+C)^2 - (A+B+C-M)^2\right]$$
$$+ 2\frac{A+B}{2}(B-M)$$
$$+ 2\left(A+B+C - \frac{A+M}{2}\right)\left(\frac{A+M}{2} - B\right) = 0.$$

Similarly, the same can be proven for point N.

Problem 23. Let $A = 0$, $G = 3B$, and $I = \frac{3D+2(3B)}{5}$, then

$$5\left[\frac{3D+2(3B)}{5} - B - D\right](D + 2B) + 3BD - 2(B^2 - D^2) = 0,$$

$$5\left(\frac{3D+2(3B)}{5} - B\right)(3B - D) - 8BD + 3(D^2 - B^2) = 0,$$

$$5\left[\left(3B - \frac{3D+2(3B)}{5}\right)(3B - D) - 3B(3B - B)\right]$$
$$+ 18BD - 3(D^2 - B^2) = 0.$$

Problem 24. Let $D = 0$, then

$$2\left(H - \frac{A+K}{2}\right)A + C(C-H) + E(C-H) + (K-A)E$$
$$+ (E-H)(A-C) + (H-K)(E-A) + (A^2 - C^2) = 0.$$

Problem 25. Let $A = 0$, $H = A + B + C$, $M = \frac{A+B+C+B}{2}$, and $E = A + B + C + B - D$: $\frac{B+A}{2}(A + B + C + B - D - B) + \frac{1}{2}\frac{B+C}{2}(B - C) + \frac{B+A}{2}(\frac{B+A}{2} - D) - \frac{A+B+C+B}{2}(\frac{A+B+C+B}{2} - D) = 0$.

Problem 26.

$$\left[\left(\frac{A+B}{2} - E\right)\left(\frac{A+B}{2} - H\right) - \left(\frac{A+B}{2} - G\right)\left(\frac{A+B}{2} - F\right)\right]$$

$$+ \frac{1}{2}(A - B)(A - E) + \frac{1}{2}(A - B)(B - F)$$

$$+ \left(\frac{A+B}{2} - F\right)(B - G) - \left(\frac{A+B}{2} - E\right)(A - H) = 0.$$

Note. For problems of this kind, setting $\frac{A+B}{2} = 0$, or $B = -A$, then simplifies the equation significantly, i.e.,

$$(EH - GF) + A(A - E) + A(-A - F) - F(-A - G) + E(A - H) = 0.$$

Problem 27. Let $O = 0$, then

$$(M - Q)P + (M - P)M - 2\frac{A+B}{2}\left(A - \frac{P+Q}{2}\right)$$

$$+ \frac{1}{2}[(A - Q)(B - P) - (Q - B)(A - P)] + A^2 - M^2 = 0.$$

Problem 28.

$$(A - E)(K - D) - (A - K)(E - K) + (A - D)(E - B)$$

$$- (A - C)(C - B) - (A - B)(C - D) + (A - K)^2 - (A - C)^2 = 0.$$

Problem 29. Let BQ intersect CR at K. Prove that K lies on AP:

$$(A - K)(F - E) + (B - K)(D - F) + (C - K)(E - D) + (A - D)(B - C)$$

$$+ (B - E)(C - A) + (C - F)(A - B) = 0.$$

Note. The identity suggests that it is sufficient for AD to be perpendicular to BC; it is not necessary for D to be on BC. The same applies to the other sides.

Problem 30.

$$4\left(\frac{A+B}{2} - \frac{A_1+B_1}{2}\right)^2 - (A-A_1)^2 - (B-B_1)^2 - 2(A-A_1)(B-B_1) = 0.$$

Problem 31. Let the centre of the circle be the origin, then

$$(O-D)(E-G) - \frac{F+B}{2}(F-C) + \frac{E+B}{2}(D-C) + \frac{F+G}{2}(F-D)$$

$$-\left(O - \frac{D+C}{2}\right)(C-D) + \left(O - \frac{D+C}{2}\right)(B-E)$$

$$-\left(O - \frac{D+F}{2}\right)(D-G) + \left(O - \frac{C+F}{2}\right)(C-B) = 0.$$

Problem 32. Let $F = 0$, then

$$2\left[\left(C - \frac{C+B}{2}\right)^2 - (E-G)\left(\frac{A}{2} - \frac{C+B}{2}\right)\right]$$

$$+ \left[(C-E)(C-A) - \left(C - \frac{C+B}{2}\right)(C-B)\right]$$

$$-2\left(\frac{C+B}{2} - A\right)(C-G) - A(C-G) - B(E-C) = 0.$$

Problem 33. Let $O = 0$, $H = A+B+C$, $N = \frac{A+B+C}{2}$, then

$$\left(\frac{A+D+E}{3} - P\right)(B-C)$$

$$+ \frac{2}{3}\left(\frac{A+B+C}{2} - A\right)(D-E)$$

$$+ \frac{1}{3}\frac{A+B}{2}(A-D) - \frac{1}{3}\frac{A+C}{2}(A-E)$$

$$+ \left(\frac{A+D}{2} - P\right)(A-B) - \left(\frac{A+E}{2} - P\right)(A-C) = 0.$$

Appendix: Exercise Answers   215

Problem 34. Let $O = 0$, $H = A + B + C$, and $N = \frac{A+B+C}{2}$, then

$$(K - X)K + \left(\frac{A+B+C}{2} - X\right)\left(\frac{A+B+C}{2} - A\right)$$

$$- \left(\frac{A+B+C}{2} - K\right)(C - X) + 2\left(\frac{A+B+C}{2} - K\right)(B - X)$$

$$+ \frac{3}{2}\left[\left(\frac{A+B+C}{2} - B\right)^2 - (K - B)^2\right]$$

$$- \frac{1}{2}\left[\left(\frac{A+B+C}{2}\right)^2 - (B + C - K)^2\right]$$

$$- 2\frac{B+C}{2}\left(\frac{B+C}{2} - X\right) = 0.$$

Note. Assuming that O is the reflection of S about BC, we have $S = B + C$, $ON = KS$, and

$$\left(\frac{A+B+C}{2}\right)^2 = (B + C - K)^2.$$

Problem 35. Let $H = 0$, then

$$4\left(F - \frac{A+E}{2}\right)\left(A - \frac{B+C}{2}\right) + 2(A - B)(A - F)$$

$$- 2F(A - C) + 2E\left(A - \frac{B+C}{2}\right) + A(B - C) = 0.$$

Problem 36. Let the circumcentre of $\triangle CEF$ be $O = 0$, then

$$2\frac{E+B}{2}\left(\frac{E+B}{2} - A\right) + (C - A)(C - B)$$

$$- \frac{1}{4}[4(C - E)^2 - (B - E)^2]$$

$$+ 3\left[C^2 - \left(\frac{E+B}{2}\right)^2\right] - (C^2 - E^2)$$

$$- 4\frac{E+C}{2}\left(C - \frac{A+B}{2}\right) = 0.$$

Problem 37. Let $O = 0$, $H = A+B+C$, $Q = \frac{A+B+C}{4}$, and the circumcentre and orthocentre of $\triangle AED$ be O_1 and H_1 respectively, with $P = \frac{O_1+H_1}{2}$. Then,

$$8(E-D)\left(\frac{A+B+C}{4} - \frac{O_1+H_1}{2}\right)$$
$$+ 3(E-C)(E-B) - 3(D-B)(D-C)$$
$$- 4(E-H_1)(A-D) + 4(D-H_1)(A-E) - 4\left(O_1 - \frac{A+E}{2}\right)(A-E)$$
$$+ 4\left(O_1 - \frac{A+D}{2}\right)(A-D) + (A-D)(A+B+C-D)$$
$$- (A-E)(A+B+C-E) = 0.$$

Problem 38. Consider the line MN intersecting the circumcircle of $\triangle CDM$ and the circumcircle of $\triangle BDN$ at points E and F, respectively. Since M is the midpoint of AB and $AD \perp BC$, we have $MD = MB$. Furthermore, $FN \parallel BD$, so M is the midpoint of FN, which implies $FM = MN$. Similarly, $NE = MN$. Let PD intersect MN at K. By the power of a point theorem, we have $KM \cdot KE = KN \cdot KF$, or $KM \cdot (KN + NE) = KN \cdot (KM + MF)$. Thus, $KM \cdot NE = KN \cdot MF$. Since $NE = MF$, it follows that $KM = KN$, which means that K is the midpoint of MN. Therefore, PD bisects MN. Alternatively, we can express this as

$$\left(K - \frac{A+C}{2}\right)\left(K - \left(2\frac{A+B}{2} - \frac{A+C}{2}\right)\right)$$
$$- \left(K - \frac{A+B}{2}\right)\left(K - \left(2\frac{A+C}{2} - \frac{A+B}{2}\right)\right) = 0,$$

which simplifies to $(B - C)\left(\frac{2A+B+C}{4} - K\right) = 0$, indicating that $K = \frac{\frac{A+B}{2} + \frac{A+C}{2}}{2}$. This problem essentially relies on the following identity:

$$\left(\frac{\frac{A+B}{2} + \frac{A+C}{2}}{2} - \frac{A+C}{2}\right)$$
$$\left(\frac{\frac{A+B}{2} + \frac{A+C}{2}}{2} - \left(2\frac{A+B}{2} - \frac{A+C}{2}\right)\right)$$
$$= \left(\frac{\frac{A+B}{2} + \frac{A+C}{2}}{2} - \frac{A+B}{2}\right)\left(\frac{\frac{A+B}{2} + \frac{A+C}{2}}{2} - \left(2\frac{A+C}{2} - \frac{A+B}{2}\right)\right).$$

Problem 39.
Note. Proving the equality of angles directly using point geometry is difficult, so a different approach is needed. Extend CF to intersect the circumcircle of $\triangle ABC$ at point K. Since $\angle BKC = \angle BAC$, the original proposition is equivalent to $BK \parallel HP$, which can be transformed into $O\frac{K+B}{2} \perp HP$. Let $O = 0$ and $H = A + B + C$. Then,

$$2\left(\frac{2F - (A+B+C) + B}{2}\right)((A+B+C) - P)$$
$$+ (C - F)(B - F) + 2F(P - F)$$
$$+ ((A + B + C) - F)(A - F) - 2\frac{A+C}{2}(P - C) = 0.$$

Problem 40.

$$[(C - E)^2 + (D - A)(D - E) - (C - D)^2] - [(C - E)^2 - (C - A)^2]$$
$$+ [(C - E)(A - D) - (C - D)(A - B)]$$
$$- [(C - A)^2 - (C - B)(C - D)] = 0.$$

Note. Since $\angle AEC = 180° - \angle ABC = \angle ACB + \angle CAB = \angle ACB + \angle CDA = \angle EAC$, we have $|CE| = |CA|$. It is easy to prove that $\triangle CED \sim \triangle ABD$.

Problem 41.
Let $C = 0$, then $P = 2B - Q$,

$$(AB' - PQ) - 2\left(P' - \frac{A}{2}\right)(A - B) + 2(P' - P)(B - Q) + (B - B')A$$
$$- [(P' - A)^2 - (P' - P)^2] = 0.$$

Note. Observing that $BQ^2 = BC' \cdot BA = BP^2$, we conclude that B is the midpoint of PQ.

Problem 42.
Let $O = 0$, then

$$P(A - C) + [(P - E)(P - H) - (P - F)(P - G)]$$
$$+ (E + H - A)(P - E) + (G + F - C)(F - P)$$
$$+ E(E - A) - F(F - C) = 0.$$

Note. Once A and C are fixed, B and D are also determined. Therefore, the relationship between B and D does not matter much in this problem and can be safely ignored in the proof.

Problem 43. Let $O = 0$, $H = A+B+C$, $N = \frac{A+B+C}{2}$, and P be a point. Then,

$$[(Q-N)(P-M) + (Q-M)(P-N)] + \frac{1}{2}P(A-B)$$

$$-\frac{1}{2}B(P-B) + \frac{3}{2}A(A-P) + [(Q-A)(Q-B) - (Q-M)(Q-N)]$$

$$-2\left(\frac{A+B}{2} - P\right)\left(\frac{A+B}{2} - Q\right)$$

$$+2\frac{M+N}{2}(P-M) - (A^2 - M^2) = 0.$$

Problem 44.

$$\left(C_1 - \frac{A+B}{2}\right)(A-B) + \left(A_1 - \frac{B+C}{2}\right)(B-C)$$

$$+ \left(B_1 - \frac{C+A}{2}\right)(C-A) - (B_1 - C_1)(A_1 - A)$$

$$- (C_1 - A_1)(B_1 - B) - (A_1 - B_1)(C_1 - C) = 0.$$

Problem 45.
Analysis. In Figure A.2, if we express the conclusion as $(D-F)^2 = (D-G)^2$, it is difficult to prove because eliminating F^2 and G^2 is challenging. So, let us prove that the symmetric point of F to D, $2D - F$, lies on BC.

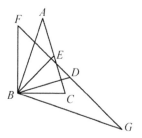

Figure A.2

Identity. Let $B = 0$, then

$$A(2D - F) - FC + 2(F - D)\frac{A+C}{2} + (C - A)D = 0.$$

Based on this identity, we can derive a more general proposition: in the diagram, in $\triangle ABC$, where $BD \perp AC$ and BE is the median, and it is given that $FD \perp BE$, $BF \perp BC$, and F is the symmetric point of D, denoted as G. We need to prove $BA \perp BG$.

Appendix: Exercise Answers

Problem 46.
Let $O = 0$, then
$$16\left(\frac{B+C+E+F}{4}\left(\frac{A+B+D+E}{4} - \frac{C+D+F+A}{4}\right)\right.$$
$$\left. - [(C-F)^2 - (B-E)^2] - 2(B^2 - C^2) + 2(F^2 - E^2)\right) = 0,$$

$$16\left(\frac{A+B+D+E}{4}\left(\frac{B+C+E+F}{4} - \frac{C+D+F+A}{4}\right)\right.$$
$$\left. - [(A-D)^2 - (B-E)^2] - 2(B^2 - D^2) + 2(A^2 - E^2)\right) = 0.$$

Problem 47.
$$2\left(\frac{X+Y}{2} - \frac{\frac{B+C}{2}+H}{2}(B-C)\right) + 2\left(X - \frac{A+C}{2}\right)\left(\frac{A+C}{2} - Y\right)$$
$$- 2\left(Y - \frac{A+B}{2}\right)\left(\frac{A+B}{2} - X\right) + (H-A)(B-C) + \frac{1}{2}(B^2 - C^2) = 0.$$

Note. To prove $\frac{X+Y}{2} - \frac{\frac{B+C}{2}+H}{2}(M-H) = 0$, we can instead prove $\frac{X+Y}{2} - \frac{\frac{B+C}{2}+H}{2}(B-C) = 0$.

Problem 48. Let $C = 0$, then
$$4\left(\frac{D}{2} - A\right)\left(\frac{D}{2} - E\right) + [(A-D)(A-B) - (E-B)B] + 2\frac{A+B}{2}(D-B)$$
$$- 2\left(A - \frac{B}{2}\right)E + [2(B-E)(A-D) - (A-D)^2] = 0,$$
$$[(D-E)^2 - D^2] + [2(B-E)(A-D) - (B-E)^2].$$

Problem 49.
Let $A = 0$, then
$$\left(kM - \frac{D}{2}\right)D + \left(kM - \frac{B+C}{2}\right)(B-C) + \frac{k-1}{2}(C^2 - (B-D)^2)$$
$$- k\left(M - \frac{B}{2}\right)B + k\left(M - \frac{C+D}{2}\right)(C-D) + (1-k)BD = 0.$$

This implies that kM lies on both the AD perpendicular line and the BC perpendicular line, which is the point N in the problem.

A.7 Exercise 5.1

Problem 1.

$$(A - O)(C - D) + (B - O)(D - E) + (C - O)(E - A) + (D - O)(A - B)$$
$$+ (E - O)(B - C) = 0.$$

Explanation. This problem might seem complex at first, but by cyclically expressing the conditions, the points naturally cancel each other out. This result is a generalization of the orthocentre theorem for triangles and can be extended to $2n+1$-gons.

Alternatively, consider a pentagon $ABCDE$, where $AF \perp CD$, $BG \perp DE$, $CH \perp AE$, and $DI \perp AB$, and they intersect at point O. Extend EO to intersect at point K. The quadrilaterals $AHFC$, $AIFD$, $BIGD$, and $OHEG$ are all cyclic. Thus, $OA \cdot OF = OH \cdot OC$, $OA \cdot OF = OI \cdot OD$, $OI \cdot OD = OB \cdot OG$, and $\angle OEG = \angle OHG$, implying $OH \cdot OC = OB \cdot OG$. Points C, B, H, and G are concyclic. From $\angle OHG = \angle OBK$, it follows that $\angle OEG = \angle OBK$. Points B, E, G, and K are concyclic, and since $BG \perp DE$, $EK \perp BC$.

Problem 2. Let $O = 0$, then

$$(A + B - K)(A - B) + (B + C - K)(B - C) + (C + A - K)(C - A) = 0.$$

A.8 Exercise 5.2

Problem 1. Consider Figure A.3.

Proof 1. Let O be the origin at the midpoint of AB. Define $E = 2(-C) - A$ and $D = 2(C + A)$. Then,

$$F = mE + (1-m)D = m[2(-C) - A] + 2(1-m)(C+A).$$

Since F lies on AB, the coefficient of C must be 0, which leads to

$$-2m + 2(1 - m) = 0,$$

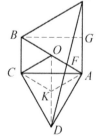

Figure A.3

and solving for m gives
$$m = \frac{1}{2},$$
so we have
$$F = \frac{1}{2}E + \frac{1}{2}D.$$

Note. The key to this problem is how to express D and E based on the assumption that A, B, and C are given. The conditions $CA \perp CB$ and $\angle B = 60°$ are not explicitly used in the solution but are used in constructing the relationships for D and E. In the diagram, G is the midpoint of AE, and K is the centre of $\triangle ACD$. With practice, you can solve it without drawing. If you note that $D + E = A$, quadrilateral $ODAE$ becomes a parallelogram with diagonals bisecting each other, making it easy to solve.

Proof 2. Let $A = 0$, $t = \frac{1+i\sqrt{3}}{2}$, $D = tC$, $E = \frac{B}{t}$, and $C = \frac{\sqrt{3}}{2}\left(\frac{\sqrt{3}}{2} + \frac{i}{2}\right)B$. Then, $\frac{D+E}{2} = \frac{B}{4}$.

Problem 2. Let $A = 0$, $B = 1$, $D = \frac{1}{2} - \frac{\sqrt{3}}{2}i$, $C = m + ni$, $C - D = \left(-\frac{1}{2} + m\right) + \left(\frac{\sqrt{3}}{2} + n\right)i$, and $F = tB + (1-t)C = (m + t - mt) + (n - nt)i$. Since $AF \parallel CD$, solving the equation $\frac{\frac{\sqrt{3}}{2}+n}{n-nt} = \frac{-\frac{1}{2}+m}{m+t-mt}$ gives $t = -\frac{\sqrt{3}m+n}{\sqrt{3}-\sqrt{3}m+n}$. Then, $F = [(2m-1) + (3+2n)i]\frac{n\sqrt{3}-3m+n}{3}$, and $E = C\left(\frac{1}{2} + \frac{\sqrt{3}}{2}i\right) = \left(\frac{m}{2} - \frac{\sqrt{3}n}{2}\right) + \left(\frac{\sqrt{3}m}{2} + \frac{n}{2}\right)i$, $B - D = \frac{1}{2} + \frac{\sqrt{3}}{2}i$, $F - E = \frac{-\sqrt{3}m+\sqrt{3}m^2+n+\sqrt{3}n^2}{2(\sqrt{3}-\sqrt{3}m+n)} + \sqrt{3}\frac{-\sqrt{3}m+\sqrt{3}m^2+n+\sqrt{3}n^2}{2(\sqrt{3}-\sqrt{3}m+n)}i$. Therefore, $EF \parallel BD$.

Problem 3.

$$[(P-R)i - (Q-R)]$$
$$- i\left[(P-C) - (B-C)\frac{\left(\cos\frac{\pi}{6} + i\sin\frac{\pi}{6}\right)\sin\frac{\pi}{4}}{\sin\left(\frac{\pi}{6} + \frac{\pi}{4}\right)}\right]$$
$$+ \left[(Q-A) - (C-A)\frac{\left(\cos\frac{\pi}{4} + i\sin\frac{\pi}{4}\right)\sin\frac{\pi}{6}}{\sin\left(\frac{\pi}{6} + \frac{\pi}{4}\right)}\right]$$
$$- (1-i)\left[(R-B) - (A-B)\frac{\left(\cos\frac{\pi}{12} + i\sin\frac{\pi}{12}\right)\sin\frac{\pi}{12}}{\sin\left(\frac{\pi}{12} + \frac{\pi}{12}\right)}\right] = 0.$$

Problem 4. Given that E, C, and F are collinear, we can assume $sF + (1-s)E - C = 0$. This leads to the following identity:

$$\left[\frac{1-2s}{1-s}A + \left(1 - \frac{1-2s}{1-s}\right)H - G\right] - \frac{i}{1-s}[sF + (1-s)E - C]$$

$$- [(A-B) - (E-B)i] + [(G-B) - (C-B)i]$$

$$+ \frac{si}{1-s}[(F-D) - (A-D)i] - \frac{si}{1-s}[(C-D) - (H-D)i] = 0.$$

Problem 5. Let $T = \cos\frac{\pi}{3} + i\sin\frac{\pi}{3}$. Then,

$$[(E_1 - A_1) - (C_1 - A_1)T] + \frac{1+\sqrt{3}i}{2}[(F_1 - B_1) - (D_1 - B_1)T]$$

$$+ \frac{1+\sqrt{3}i}{2}[(C_1 - B) - (A - B)T] + \frac{-1+\sqrt{3}i}{2}[(D_1 - C) - (B - C)T]$$

$$- [(E_1 - D) - (C - D)T] - \frac{1+\sqrt{3}i}{2}[(F_1 - E) - (D - E)T]$$

$$+ \frac{1-\sqrt{3}i}{2}[(A_1 - F) - (E - F)T] + [(B_1 - A) - (F - A)T] = 0.$$

Note. This implies that in triangles $\triangle ABC_1$, $\triangle BCD_1$, $\triangle CDE_1$, $\triangle DEF_1$, $\triangle EFA_1$, $\triangle FAB_1$, $\triangle B_1D_1F_1$, and $\triangle A_1C_1E_1$, any seven of them are equilateral triangles, and the remaining one must also be an equilateral triangle.

Problem 6. Let $T = \frac{1+i}{2}$, then $\left[\left(\frac{D_1+A_1}{2} - \frac{B_1+A_1}{2}\right) - \left(\frac{B_1+C_1}{2} - \frac{B_1+A_1}{2}\right)i\right] - \frac{i}{2}[(A_1 - B) - (A - B)T] + \frac{1}{2}[(B_1 - C) - (B - C)T] + \frac{i}{2}[(C_1 - D) - (C - D)T] - \frac{1}{2}[(D_1 - A) - (D - A)T] = 0.$

Problem 7. Let $T = \frac{1+i}{2}$, then

$$(D_1 - A_1) - (B_1 - A_1)i + (1-i)\left[\left(A_1 - \frac{B+C}{2}\right) - \left(\frac{A+B}{2} - \frac{B+C}{2}\right)T\right]$$

$$+ i\left[\left(B_1 - \frac{C+D}{2}\right) - \left(\frac{B+C}{2} - \frac{C+D}{2}\right)T\right]$$

$$- \left[\left(D_1 - \frac{A+B}{2}\right) - \left(\frac{D+A}{2} - \frac{A+B}{2}\right)T\right] = 0.$$

A.9 Exercise 5.3

Problem 1. In Figure A.4, $2\left(\frac{C+D}{2} - P\right)$
$(A - B) - (C - A)(C - B) + (D - A)$
$(D - B) + [(A - C)^2 - (A - P)^2] - [(B - D)^2 - (B - P)^2] = 0$.

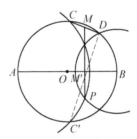

Figure A.4

Note. The original question emphasizes the semicircle $\odot O$, with C and D on the same side of AB and P on the diameter AB; however, from the identity, these emphases are not necessary. Generalization: given a point P inside the circle O with AB as the diameter. Draw circle A with centre A and radius AP. Circle A intersects circle O at point C. With point B as the centre and BP as the radius, draw circle B. Circle B intersects circle O at point D, and the midpoint of the line segment CD is M. Prove that MP is perpendicular to AB.

A.10 Exercise 5.4

Problem 1. Let $O = 0$, then

$$[2PA - (A^2 + B^2)] + 2B(B - P) + [(P - A)^2 - (P - B)^2] = 0.$$

The locus is a straight line perpendicular to OA and satisfies $2\overrightarrow{OP} \cdot \overrightarrow{OA} = OA^2 + OB^2$.

Problem 2. Let $C = 0$, then

$$\left[\frac{A + B + C}{3} - (xB + (1 - x)C)\right]^2 - y(B - C)^2$$

$$+ k\left[\left(B - \frac{A + C}{2}\right)^2 - 4\left(C - \frac{A + B}{2}\right)^2\right] = 0,$$

i.e.,

$$\frac{1}{36}(4 - 27k)A^2 + \frac{1}{9}(2 - 27k - 6x)AB + \frac{1}{9}(1 - 6x + 9x^2 - 9y)B^2 = 0,$$

Solving the system of equations:

$$4 - 27k = 2 - 27k - 6x = 1 - 6x + 9x^2 - 9y = 0,$$

we get $k = \frac{4}{27}$, $x = -\frac{1}{3}$, $y = \frac{4}{9}$, and obtain the identity

$$\left(\frac{A+B+C}{3} - \left(-\frac{1}{3}B + \frac{4}{3}C\right)\right)^2 - \frac{4}{9}(B-C)^2$$

$$+ \frac{4}{27}\left(\left(B - \frac{A+C}{2}\right)^2 - 4\left(C - \frac{A+B}{2}\right)^2\right) = 0,$$

which shows that the locus of point P is a circle with center $-\frac{1}{3}B + \frac{4}{3}C$ and radius $\frac{2}{3}BC$.

Note. Although this problem seems to test geometry, it can also be viewed as a test of algebraic transformation. Rewrite the condition $\left(B - \frac{A+C}{2}\right)^2 - 4\left(C - \frac{A+B}{2}\right)^2 = 0$ into the form of (moving point P − center)^2 = radius^2.

www.ingramcontent.com/pod-product-compliance
Lightning Source LLC
LaVergne TN
LVHW022313291224
800089LV00002B/42